Shape and Number 4

R R Joy B.Sc., M.Ed.

Macmillan Education

First published 1976
Reprinted 1978, 1979

Published by
Macmillan Education Limited
Houndmills Basingstoke Hampshire RG21 2XS
and London
Associated companies in Delhi Dublin
Hong Kong Johannesburg Lagos Melbourne
New York Singapore and Tokyo

Filmset in Great Britain by BAS Printers Ltd, Wallop, Hampshire
and printed in Hong Kong by
The Hong Kong Printing Press (1977) Ltd

Contents

Section B

Section C

Section H

Section I

Preface

This series is intended for use by children of a wide range of mathematical ability. Both content and approach contain a mixture of the traditional and the modern, and the final book contains all the work required for the core of a traditional type of 16+ examination, including the modern topics most commonly incorporated into mode 3 syllabuses.

Each book is divided into sections covering approximately half a term's work. Rather than a prolonged study of any one area, each section contains a number of short chapters covering different topics. Thus, by making short strides forward and returning to the same area of study several times each year, children are encouraged to keep up with the pace and yet to feel that they are on familiar ground. No topic is considered in isolation, since this arrangement facilitates frequent cross referencing.

In order to give teachers flexibility in the choice of material, each book contains a number of well-graded examples giving practice in algebraic manipulation and computational skills. It is not intended that any child should be asked to work systematically through all the examples which are provided. There are also many worked examples with step-by-step explanations.

This fourth book covers the final two years' work of a five-year course, but because it contains all the work required for examination purposes, it is equally well suited to an Upper school, where the children may come from several different Middle schools with a variety of mathematical backgrounds.

The work is arranged over four-and-a-half terms, leaving time for the study of special topics in depth, and for revision. Each chapter contains two exercises. The first of these exercises consists of short examples which give practice in basic skills, while the second contains a set of multi-facet questions of examination standard, which can be used either as the course progresses, or at the end for revision purposes.

The use of four-figure tables and the slide rule are explained using programmes, which are intended to overcome some of the difficulties inherent in teaching these particular skills. It is suggested that children should work through the relevant programme after basic instruction has been given.

Schools preparing children for examinations for which the use of four-figure tables is not required, will find that most questions are suitable for use with three-figure tables, although answers are given correct to three significant figures or to the nearest minute. Cosines and cosine tables are explained, but apart from examples giving practice in the use of tables, cosines are not essential for answering the questions, which are restricted to simple problems requiring the solution of right-angled triangles.

The intention is to present familiar material with clarity, and in a way which is appropriate to its users. It is hoped that it will lend support to non-specialist teachers without in any way restricting the ingenuity of specialist mathematicians.

I would like to thank Tony Feldman and Tim Pridgeon, and all those who have contributed to the preparation of this series.

Leeds 1976 R. R. Joy

Section A
1 Sets

Sets and Elements

A set is a collection. Sets are usually shown by capital letters. For example:

D = {days of the week}
or S = {stars in the Milky Way}.

We can talk of a set by describing its members, as above, or by listing them. The members of a set are called its *elements*. If we list the elements of set D, we have:

D = {Sunday, Monday, Tuesday, Wednesday, Thursday, Friday, Saturday}.

Just as sets are represented by capital letters, individual elements are usually represented by small letters. We can say that:

$s \in D$ (Sunday is an element of D)
$m \in D$ (Monday is an element of D)

and so on. Also,

April $\notin D$ (April is not an element of D)
Neptune $\notin D$ (Neptune is not an element of D).

If we try to list the elements of S we find that it cannot be done. We do not know of all the stars in the Milky Way, and there are not names for all those we do know. In any case, there are far too many of them. It is much easier to say that:

S = {stars in the Milky Way}.

Union and Intersection

Some elements may be in more than one set. Suppose,

P = {boys in your class}
and Q = {blue-eyed boys}.

There is probably at least one blue-eyed boy in your class, and he will be in both sets. This can be represented on a diagram as shown.

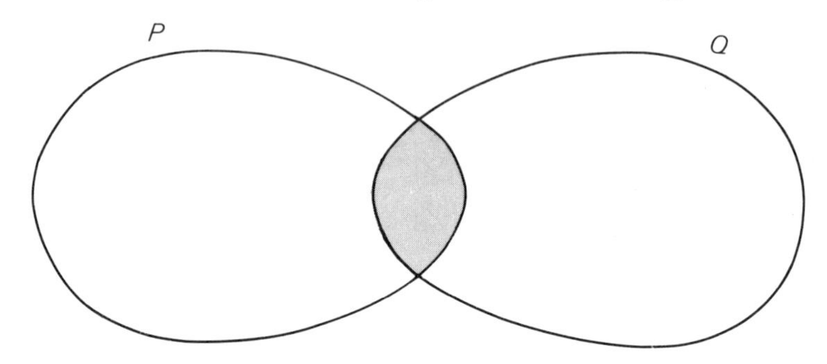

A diagram like this is called a *Venn diagram*. The shaded part shows where the sets overlap, and is called the *intersection* of the sets. It is written $P \cap Q$ and spoken as P intersect Q. It contains all the elements which are in both P and Q. In this case it represents all blue-eyed boys in your class. Shading the whole diagram shows all the elements in the two sets. It shows all the elements which are either in P or in Q. It is called the *union* of P and Q. It is written as $P \cup Q$ and spoken as P union Q.

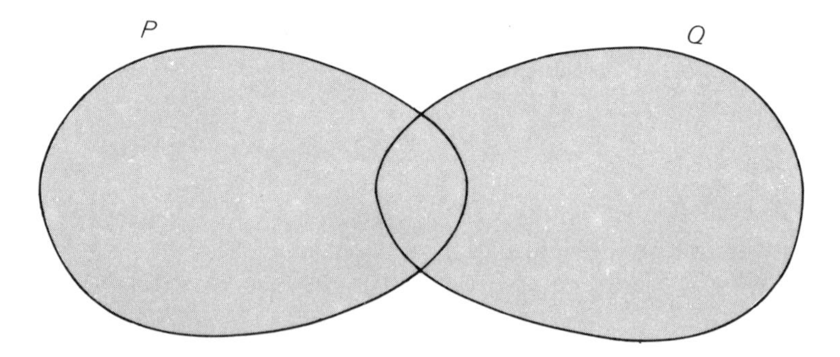

Subsets

Sometimes the whole of a set is part of a larger set. Suppose,

E = {towns in England}
G = {towns in Great Britain}.

All the elements of E are also elements of G. E is called a *subset* of G, and we write $E \subset G$ (E is a subset of G).

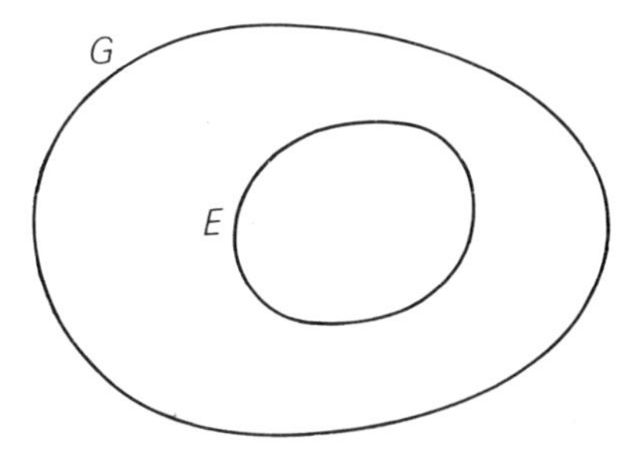

In this case

$$E \cap G = E$$
and $E \cup G = G$.

Envelope and Complement

The sets shown on page 2 are:

P = {boys in your class}
Q = {blue-eyed boys}.

Each of these sets is a subset of a set which contains both P and Q. If B = {boys}, we have $P \subset B$ and $Q \subset B$. B contains all the other sets and is called the universal set, or *envelope*. It is represented on a Venn diagram by a rectangle, and the letter $\mathscr{E}$.

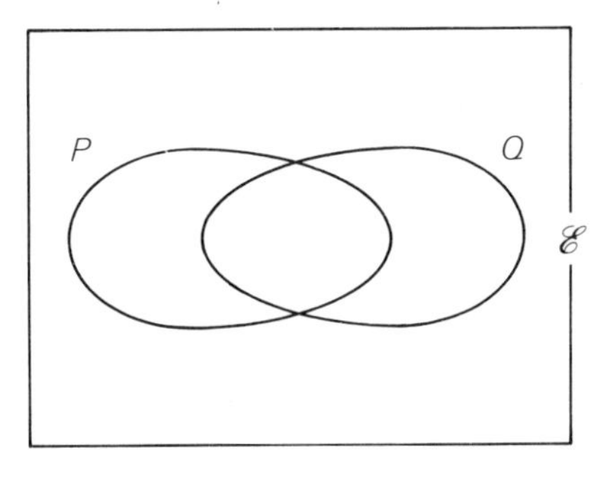

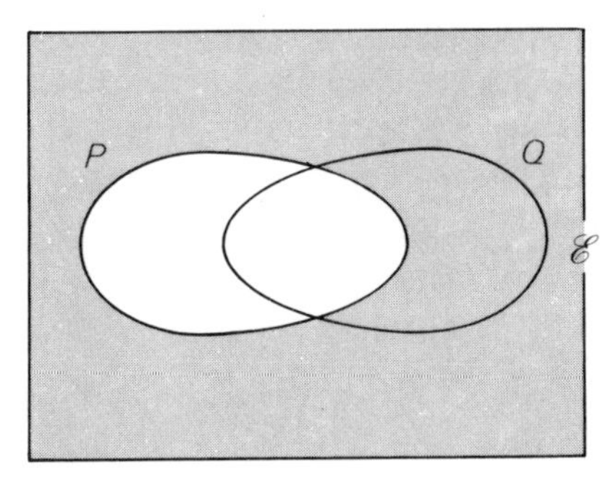

In the shaded diagram, the shaded part shows the elements in $\mathscr{E}$ which are not in P. This shaded part is called the *complement* of P, and is written as P'. Thus,

P = {boys in your class}
P' = {boys not in your class}
and $P \cup P' = \mathscr{E}$.

Cardinal Number

$n(A)$ means the number of elements in set A. If $D = \{\text{days of the week}\}$, then $n(D) = 7$. The number of elements in the set is called the *cardinal number* of the set.
Consider the sets,

$L = \{\text{icebergs on the sun}\}$
$M = \{\text{cities on the moon}\}$.

We have $n(L) = 0$, and $n(M) = 0$. A set which has no elements in it is called an *empty set*, and is shown by the Greek letter ϕ (pronounced fi as in find).

Two sets which contain exactly the same elements are called *equal sets.*

Two sets S and T in which $n(S) = n(T)$, are called *equivalent sets.*

Finite and Infinite Sets

A set such as {days of the week} or {boys in your class}, which has a finite number of elements is called a *finite set.*

In mathematics we are often concerned with sets of numbers. Consider, for example, the set

$N = \{1, 2, 3, 4, 5, 6, \ldots\}$.

This is called the set of *natural numbers*. However many numbers you write in the brackets, you can never get to the end of the set. The numbers just go on and on. This is shown by the three dots. There are an infinite number of elements in this set. A set with an infinite number of elements is called an *infinite set.*

Exercise 1a

If $L = \{\text{letters of the alphabet}\}$, $G = \{\text{Greek letters}\}$ and $P = \{\text{numbers from 1 to 10}\}$, say whether the following statements are true or false.

(1) $6 \in P$
(2) $3 \notin L$
(3) $x \in P$
(4) $\pi \in G$
(5) $\pi \notin L$.

Look at the Venn diagrams and say whether the following statements are true or false.

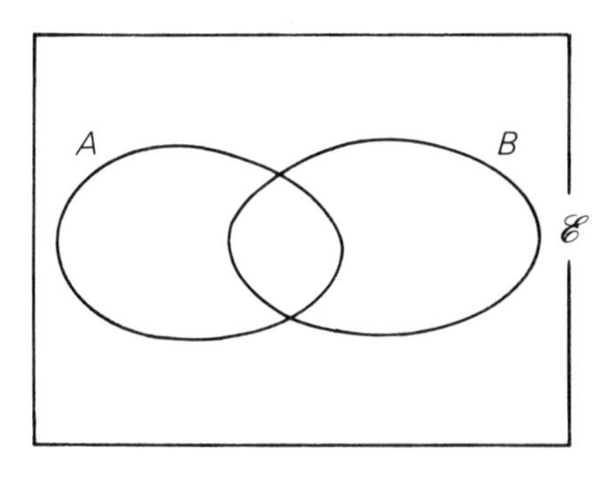

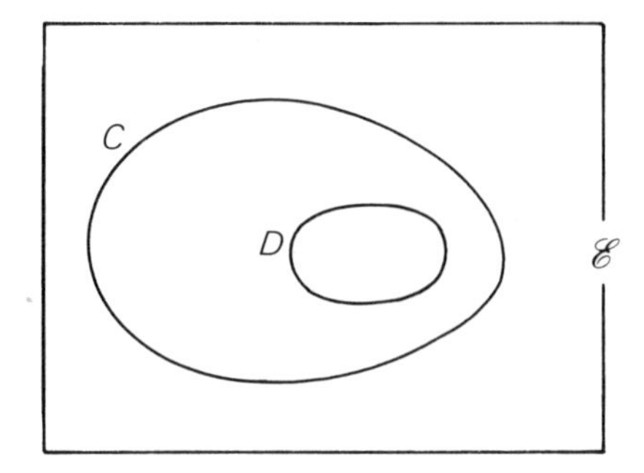

(6) $A \subset B$
(7) $B \subset \mathscr{E}$
(8) $A \cup B = \phi$
(9) $C \subset D$
(10) $C \cup D = D$.

(11) Copy the Venn diagram above, and shade in $A \cap B$.
(12) Copy the Venn diagram above, and shade in $A \cup B$.
(13) Copy the Venn diagram above, and shade in B'.
(14) Copy the Venn diagram above, and shade in A'.
(15) Copy the Venn diagram above, and shade in $\mathscr{E}$.
(16) Copy the Venn diagram above, and shade in $(A \cap B)'$.
(17) Copy the Venn diagram above, and shade in $A' \cap B$.
(18) Copy the Venn diagram above, and shade in $C \cap D$.
(19) Copy the Venn diagram above, and shade in $C \cup D$.
(20) Copy the Venn diagram above, and shade in C'.
(21) If $R = \{\text{red-heads}\}$, and $I = \{\text{Irishmen}\}$, what does $R \cap I$ represent?
(22) If $W = \{\text{women}\}$, and $A = \{\text{athletes}\}$, what does $W \cap A$ represent?
(23) If $B = \{\text{bow-legged people}\}$, and $J = \{\text{jockeys}\}$, what does $B \cap J$ represent?
(24) If $A = \{\text{absent-minded people}\}$, and $P = \{\text{professors}\}$, what does $A \cap P$ represent?
(25) If $W = \{\text{white animals}\}$, and $R = \{\text{rabbits}\}$, what does $W \cap R$ represent?
(26) If $M = \{\text{men}\}$, and $W = \{\text{women and children}\}$, what does $M \cup W$ represent?
(27) If $P = \{\text{schoolmasters}\}$, and $Q = \{\text{schoolmistresses}\}$, what does $P \cup Q$ represent?
(28) If $K = \{\text{knives}\}$, and $F = \{\text{forks and spoons}\}$, what does $K \cup F$ represent?

(29) If S = {the sun}, and P = {planets}, what does $S \cup P$ represent?
(30) If A = {arms}, and L = {legs}, what does $A \cup L$ represent?

Write down the cardinal numbers asked for in questions 31–40.

(31) $n(M)$, where M = {months}.
(32) $n(S)$, where S = {sides of triangle ABC}.
(33) $n(C)$, where C = {pieces of a chess set}.
(34) $n(F)$, where F = {members of a football team}.
(35) $n(Y)$, where Y = {years in a century}.
(36) $n(W)$, where W = {wheels on a bicycle}.
(37) $n(D)$, where D = {days in September}.
(38) $n(P)$, where P = {playing cards in a pack}.
(39) $n(L)$, where L = {days in a leap year}.
(40) $n(M)$, where M = {pennies in a pound}.
(41) Which of the sets given in questions 31–40 are equivalent sets?
(42) Which of the sets in questions 31–40 is equivalent to {socks in a pair}?
(43) Which of the sets in questions 31–40 is equivalent to {signs of the zodiac}?
(44) Which of the sets in questions 31–40 is equivalent to {months in the first quarter of the year}?
(45) Which of the sets in questions 31–40 is equivalent to {a brace of pheasants}?

In questions 46–50 say which of the sets are finite and which are infinite.

(46) D = {days in the year}.
(47) N = {the natural numbers}.
(48) F = {fractions}.
(49) E = {people in England}.
(50) P = {points on a line}.

Exercise 1b

(1) Consider the sets $\mathscr{E}$ = {people}, B = {blind people}, D = {deaf people}.
   (a) Draw a Venn diagram to illustrate these sets.
   (b) On your diagram shade $B \cap D$.
   (c) What does $(B \cup D)'$ represent?
   (d) What does $D' \cap B$ represent?
(2) Consider the sets $\mathscr{E}$ = {numbers}, N = {natural numbers}, E = {even numbers}.
   (a) List the first five elements of N.

(b) Draw a Venn diagram to illustrate these sets.
(c) On your diagram shade N'.
(d) List the first five elements of $E \cap N$.

(3) $\mathscr{E}$ = {boys and girls}, A = {blue-eyed}, B = {fair-haired}, C = {girls}.
(a) Draw a Venn diagram to illustrate these sets.
(b) Shade $A \cap B \cap C$.
(c) What does $C' \cap A$ represent?
(d) What does $B \cap C \cap A'$ represent?

(4) In the sixth form of Hilltop School, pupils have a choice of three science subjects. P = {pupils taking Physics}, C = {pupils taking Chemistry}, B = {pupils taking Biology}.
(a) Draw a Venn diagram to illustrate these sets.
(b) What does $P \cap C \cap B$ represent?
(c) On your diagram shade $(P \cup C) \cap B$.
(d) If $n(P) = 6$, $n(C) = 5$, $n(P \cup C) = 8$, how many pupils take both Physics and Chemistry?

(5) In the sixth form of Hilltop School, eight students take English, and ten take French.
(a) If there are 15 students taking these subjects, how many of them take both?
(b) If 2 of the English students also take History,
3 of the French students also take History,
1 student takes only History,
no student takes all three subjects,
illustrate the sets on a Venn diagram, and show the number of students in each region of the diagram.
(c) What is the total number of students taking one or more of these three subjects?
(d) If E = {English students}, F = {French students}, H = {History students}, what is the value of $n[(E \cup H) \cap F]$?

2 Natural Numbers

Numerals

The figures we use, 1, 2, 3, 4, 5, etc., are called *numerals*. They have come to us from India, through the Arab world. The great advantage of this *Hindu-Arabic notation* is that the value given to any one of the symbols depends on its position.
Look at the number 8 088.

Thousands	Hundreds	Tens	Units
8	0	8	8

The number 8 is written three times, yet each 8 means something different. Starting from the right, there are eight units, eight tens, and eight thousands. The zero keeps a space between the tens column and the thousands column. Without it, the number would look like eight hundred and eighty-eight. Without a zero, the number 650 would look like 65.

Just think how much easier it is to write 3 768, than to use Roman numerals, and write MMMDCCLXVIII. The Romans did not have a symbol for zero, and as a result, they were very bad at arithmetic.

Even when using our Hindu-Arabic numerals, it is easy to make mistakes if you do not keep to the rules. A few examples will show the way in which the rules work. The names of the rules are given, but these are not important. The rules are:

Commutative Law

This is a very simple law. Let us think of the numbers 3 and 6. The law tells us that:

$$3 + 6 = 6 + 3$$

and $3 \times 6 = 6 \times 3$.

Is $6 - 3 = 3 - 6$?
Is $6 \div 3 = 3 \div 6$?

Distributive Law

This law is used when addition and multiplication are being carried out together. For example,

$2 \times (3 + 5)$ means $2 \times 8 = 16$ .... (1)

The law tells us that

$2 \times (3 + 5) = 2 \times 3 + 2 \times 5 = 6 + 10 = 16$ .... (2)

Notice that in (1) the value of the numbers inside the brackets is worked out first. In (2), the multiplications have been carried out before the additions. This is very important.

Associative Law

First let us look at addition. Suppose we are adding three numbers. This law tells us that it does not matter which two we add first. For example,

$(3 + 7) + 5 = 10 + 5 = 15$
$3 + (7 + 5) = 3 + 12 = 15.$

Also, because the order is not important,

$3 + (5 + 7) = 3 + 12 = 15$
$(3 + 5) + 7 = 8 + 7 = 15.$

The same is true for multiplication:

$(2 \times 3) \times 4 = 6 \times 4 = 24$
$(2 \times 4) \times 3 = 8 \times 3 = 24$
$(3 \times 4) \times 2 = 12 \times 2 = 24.$

In each of these examples, brackets have been used to show which pair of numbers have been grouped together.

Is $24 \div (6 \div 2)$ the same as $(24 \div 6) \div 2$?

The formula for the area of a triangle should be familiar. It is,

area $= \frac{1}{2}bh.$

Does this mean (half of the base) $\times$ (height) or half of (base $\times$ height)? The answer is that it means both of these things. It also means (half of the height) $\times$ base. Thus, if the base is 4 units and the height is 6 units, then:

area $= \frac{1}{2}bh = \frac{1}{2} \times 4 \times 6$.

This can be worked out in several different ways.

$(\frac{1}{2} \times 4) \times 6 = 2 \times 6 = 12$ square units
$(\frac{1}{2} \times 6) \times 4 = 3 \times 4 = 12$ square units
$\frac{1}{2} \times (4 \times 6) = \frac{1}{2} \times 24 = 12$ square units

$\frac{4 \times 6}{2} = \frac{24}{2} = 12$ square units

$\frac{\overset{2}{\cancel{4}} \times 6}{\underset{1}{\cancel{2}}} = 2 \times 6 = 12$ square units

$\frac{4 \times \overset{3}{\cancel{6}}}{\underset{1}{\cancel{2}}} = 4 \times 3 = 12$ square units.

The last two ways use the method of cancelling. More will be said about this later.

Exercise 2a

Find the value of the following:

(1) $3(5 + 2)$
(2) $7 \times 5 + 6$
(3) $3 + 8 \times 9$
(4) $3 + 2 \times 2 + 3$
(5) $3 \times 2 + 2 \times 3$
(6) $\frac{1}{2} \times 3 \times 4$
(7) $3 \times 4 \div 2$
(8) $5 \times 2 \times 3$
(9) $3 \times (6 \div 2)$
(10) $(20 \div 2) \div 2$
(11) $40 \div (8 \div 4)$
(12) $(3 + 7)(5 - 4)$
(13) $10 - 3 \times 2$
(14) $10 \times 2 - 3$
(15) $\frac{7 \times 6}{2}$.

Say whether the following statements are true or false without working out the values.

(16) $632 + 583 = 583 + 632$
(17) $764 - 243 = 243 - 764$
(18) $346 \times 213 = 213 \times 346$
(19) $76 \times 45 = 45 \times 70 + 45 \times 6$
(20) $653 \div 212 = 212 \div 653$
(21) $42(16 + 73) = 42 \times 16 + 42 \times 73$
(22) $54 \times 33 + 22 \times 33 = 33(54 + 22)$
(23) $(71 + 32) + 10 = 71 + (32 + 10)$
(24) $54 + 76 = 56 + 74$
(25) $60 \div (6 \div 2) = (60 \div 6) \div 2$

(26) $\frac{1}{2} \times 8 \times 4 = \frac{8}{2} \times \frac{4}{2}$
(27) $(54 \times 33) \times 21 = 54 \times (33 \times 21)$
(28) $(45 \times 24) \div (8 \times 7) = 45 \times (24 \div 8) \times 7$
(29) $33(45 - 17) = 33 \times 45 - 33 \times 17$
(30) $37 \times (14 - 2) = (37 \times 14) - 2$

Simplify the following. Do not work out the value.

(31) $23 \times 14 + 17 \times 14$
(32) $15 \times 20 - 7 \times 20$
(33) $57 \times 23 - 57 \times 15$
(34) $42 \times 3 + 42 \times 11 + 42 \times 2$

Work out the following:

(35) 11 487 + 1 612 + 793 + 2 337 + 833 + 10 121
(36) 48×87 (check by finding 87×48 and $87 \times 6 \times 8$)
(37) 72×163 (check by finding 163×72 and $163 \times 6 \times 12$)
(38) $2\,208 \div 23$
(39) $1\,540 \div 35$
(40) $12\,033 \div 21$
(41) $45 \times 4\,000$
(42) $4\,840 \div 20$
(43) $43 \times 50 + 17 \times 50$
(44) $117 \times 20 - 17 \times 20$
(45) $(45 \times 30) \div (15 \times 30)$.

Exercise 2b

(1) A milkman delivers an average of two bottles of milk to each house.

(a) How many bottles does he deliver altogether, if he calls at 572 houses?

(b) If each crate holds 16 bottles, how many crates must he have on the van?

(c) When he has called at 496 houses, how many bottles are left in the van?

(2) A bricklayer finds that he can lay an average of 729 bricks each day.

(a) Calculate the number of bricks laid in a six-day week.

(b) If he lays 3 962 bricks from Monday to Friday, how many must he lay on Saturday to keep up his average?

(c) How many bricks does he lay in a month if he works 27 days?

(d) How many days must he work in order to lay 10 000 bricks?

(3) A man drives a journey in a car with an average speed of 55 km/h, and an average fuel consumption of 11 km per litre.

(a) If the journey takes seven hours, how far does the man drive?

(b) How much fuel is used?

(c) What is the rate at which petrol is being used?

(d) If there is just enough fuel in the tank for the complete journey, how much is left after travelling 275 km?

(4) A theatre has 20 rows of seats, with 16 seats to each row. Seats in the first three rows cost £2. All other seats are £1 each.

(a) How many seats are there in the theatre?

(b) What is the maximum number of tickets available for one week, if there is one performance each evening from Monday to Saturday, and also a matinee performance each Wednesday and Saturday afternoon?

(c) Calculate the income from sales of tickets for one performance, if all the seats are booked.

(5) A football team has eleven players.

(a) How many players take part in 27 matches if each team plays only one game?

(b) If each team has four reserves, how many reserves must there be for these 27 matches?

(c) What is the total number of players and reserves involved?

(d) How many rugby players are there in 54 teams? (15 players in each team.)

3 Simple Algebra

Letters for Numbers

In Algebra, letters are used instead of numbers. If, for example, a chocolate bar costs 5 pence, then n bars will cost $5n$ pence. Writing C for cost, we have:

$C = 5n$ pence.

This is a *formula* for finding the cost of any number of bars. In order to find the cost of ten bars, we must *substitute* the value $n = 10$ into the formula. The cost of ten bars is $C = 5 \times 10 = 50$ pence.

Adding and Multiplying

2 fours + 3 fours = 5 fours.

In the same way

$$2x + 3x = 5x.$$

The number in front of the x, that is 2, 3 or 5, is called the *coefficient* of x.

Just as $x + x$ is shortened to $2x$,
$x \times x$ is shortened to $x \cdot x$ or x^2 (x squared).

The little 2, which shows that two x's are multiplied together is called the *power* of x, or the *index* (say indices for more than one index).

If three or more x's are multiplied together, we have:

$$x \cdot x \cdot x = x^3 \text{ (x cubed)}$$
$$x \cdot x \cdot x \cdot x = x^4 \text{ (x to the fourth)}$$
$$x \cdot x \cdot x \cdot x \cdot x = x^5 \text{ (x to the fifth)} \quad \text{and so on.}$$

Expressions

Something like $4x + 5x^2$ is called an algebraic *expression*. $4x$ is the first *term* of the expression. $5x^2$ is the second term. The coefficient of x is 4. The coefficient of x^2 is 5. Note that $5x^2$ means $5 \times x \times x$. It is only the x which is squared, not the 5.

In order to find the value of an expression, such as $4x + 5x^2$, we must substitute the correct value of x. Thus, if x is 3,

$$\begin{aligned} 4x + 5x^2 &= (4 \times 3) + (5 \times 3 \times 3) \\ &= 12 + 45 \\ &= 57. \end{aligned}$$

Laws of Algebra

The letters which we use in algebra stand for numbers. They must therefore obey the laws of numbers which were given in chapter 2. Thus, for two numbers a and b,

$$a + b = b + a$$

and $$ab = ba.$$

Note that $a \times b$ is usually written as ab, without the multiplication sign.

If three letters a, b and c are to be added, it does not matter which two are added first. If they are to be multiplied, it does not matter which two are multiplied first.

Also,

$$a(b + c) = ab + ac.$$

When finding the value of an expression which has brackets in it, do not try to split it up, but find the value of the brackets before doing anything else.

Example: Find the value of $3(1 + x)^2$, when $x = 5$.

$$\begin{aligned} 3(1 + x)^2 &= 3 \times 6^2 \\ &= 3 \times 36 \\ &= 108. \end{aligned}$$

Again, notice that it is only the 6 which is squared, not the 3.

Exercise 3a

(1) How many days are there in x weeks?
(2) How many hours are there in y days?
(3) How many months are there in z years?
(4) What is the cost of n records at 70 pence each?
(5) What is the total weight of x bags of flour, each weighing 2 kg?
(6) If n is an even number, what is the next even number?
(7) A car travels at 30 km per hour. How far does it travel in x hours?
(8) A car travels 10 km to each litre of petrol. How far can it travel on n litres?
(9) What is the cost of n metres of cloth at p pence per metre?
(10) If tea is made by using one spoonful of tea per person and one for the pot, how many spoonfuls are needed for x people?

In questions 11–20, write the expressions without brackets.

(11) $3(x + y)$
(12) $2(4x + y)$
(13) $5(p - q)$
(14) $3(2p - 3q)$
(15) $7(x^2 + 1)$
(16) $5(2 - 3x^2)$
(17) $2x^2(y + z)$
(18) $x^3(1 + y)$
(19) $p(x - y^2)$
(20) $q(x^2 + 3y)$

In questions 21–25 say whether the statements are true or false.

(21) $x + y = y + x$
(22) $pq = qp$
(23) $3x^2 = 3x \times 3x$
(24) $4x^3 = 4 \times x \times x \times x$
(25) $x^2 + y^2 = (x + y)^2$

Find the values of the following expressions:

(26) $2x + 3y$, when $x = 4, y = 3$
(27) $5a - 2b$, when $a = 3, b = 1$
(28) $3xy$, when $x = 7, y = 2$
(29) $2ab$, when $a = 2, b = 3$
(30) $4x^2$, when $x = 3$
(31) $2x^3$, when $x = 2$
(32) abc, when $a = 1, b = 2, c = 3$
(33) $\frac{xy}{z}$, when $x = 2, y = 6, z = 4$
(34) $a^2 + b^2$, when $a = 3, b = 2$
(35) $3(x - y)$, when $x = 5, y = 4$
(36) $4(a + b)$, when $a = 7, b = 2$
(37) $5(y - z)$, when $y = 5, z = 0$
(38) $7(p - q)$, when $p = 11, q = 10$
(39) $3(x^2 - y^2)$, when $x = 5, y = 1$

(40) $2(a^2 + b^2)$, when $a = 3, b = 4$
(41) $2x(a + b)$, when $x = 1, a = 2, b = 3$
(42) $2ax + 2bx$, when $x = 1, a = 2, b = 3$
(43) $x^2(a - b)$, when $x = 3, a = 7, b = 4$
(44) $(x + y)^2$, when $x = 3, y = 4$
(45) $(2x - y)^2$, when $x = 2, y = 3$
(46) $(x + y) + (x + y)^2$, when $x = 1, y = 2$
(47) $3xy^2$, when $x = 5, y = 0$
(48) $x(y + z)$, when $x = 0$
(49) $5(x^2 + y^2)$, when $x = 1, y = 2$
(50) $5(x - y)^2$, when $x = 4, y = 3$.

Exercise 3b

(1) A bicycle wheel has a diameter $d = 70$ cm. Taking π as $\frac{22}{7}$, calculate:
(a) the circumference of the wheel in cm ($C = \pi d$),
(b) the distance travelled in metres for fifty complete turns of the wheel,
(c) the number of complete turns necessary to travel 1 km.

(2)

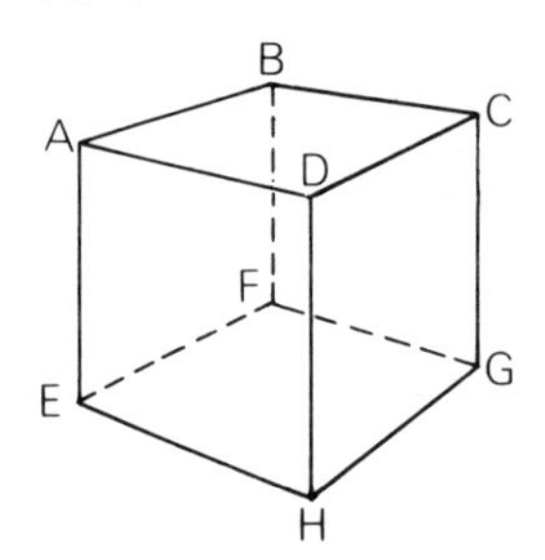

A cube ABCDEFGH as shown has edges $l = 4$ cm long. Calculate:
(a) the total length of all the edges (AB + BC + CD etc.),
(b) the area of each face in square centimetres ($A = l^2$),
(c) the total surface area of the cube in cm² ($S = 6l^2$),
(d) the volume of the cube in cm³ ($V = l^3$).

(3) A sphere has radius $r = 7$ cm. Taking π as $\frac{22}{7}$, calculate:
(a) the area of its cross-section in cm² ($A = \pi r^2$),
(b) the area of its surface in cm² ($S = 4\pi r^2$),
(c) the volume of the sphere in cm³ ($V = \frac{4}{3}\pi r^3$).

(4) A stone is thrown upward from ground level with a speed of 30 metres per second.
(a) The formula $v = 30 - 10t$ gives the velocity v of the stone in m/s after t seconds. Find the velocity after 2 seconds.

(b) Use your answer to (a) in the formula $h = \frac{(30 + v)t}{2}$ to find the height of the stone.

(c) The formula $h = 30t - 5t^2$ gives the height of the stone in metres after t seconds. Use the formula to find the height of the stone after 4 seconds. Also, use this formula to check your answer to (b).

(5) If $x = 6t$ and $y = 3t^2$,

(a) find x when $t = 5$,

(b) find y when $t = 5$,

(c) show that when $t = 5$, $x^2 = 12y$.

4 Decimals

Adding and Subtracting

Adding and subtracting decimals is much the same as adding and subtracting natural numbers.

The following two examples show this.

Example: 3 597 + 295 + 36 + 841	Example: 35.97 + 2.95 + 0.36 + 8.41
3 597	35.97
295	2.95
36	0.36
841	8.41
4 769	47.69

In each case it is important that units are below units, tens are below tens, and so on. If the sum is set out correctly there should be no difficulty.

The same applies to subtraction.

Example: 5 976 − 385	Example: 59.76 − 3.85
5 976	59.76
− 385	− 3.85
5 591	55.91

If an example contains both addition and subtraction, these must be done separately.

Example: 273.1 − 52.41 + 38.4 − 97.3

273.1	52.41	311.50
+ 38.4	+ 97.3	− 149.71
311.5	149.71	161.79

1. Add the items to be added.

2. Add the items to be subtracted.

3. Subtract total 2 from total 1.

Multiplying

Multiplication of decimals is done by writing the numbers as fractions. To multiply 37.5 × 4.36 we say that

$$\begin{aligned} 37.5 \times 4.36 &= \frac{375}{10} \times \frac{436}{100} \\ &= \frac{163\,500}{1\,000} \\ &= 163.500. \end{aligned}$$

This can be simplified as follows:

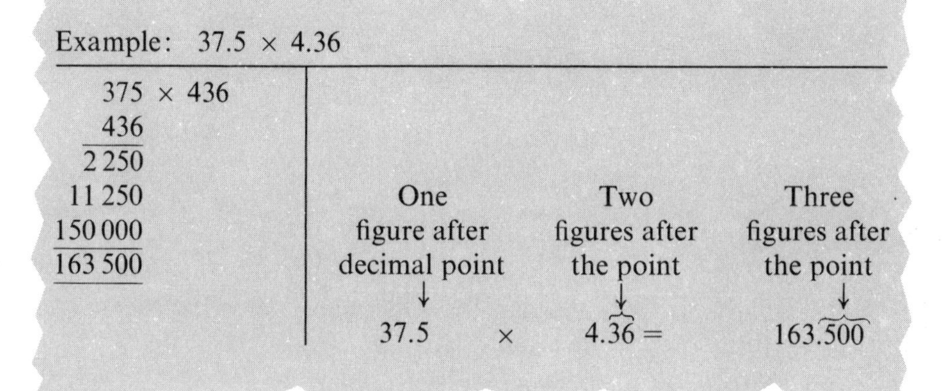

1. Multiply the numbers without the decimal points.

2. Replace the decimal points. The number of figures after the point in the answer must be the same as the total number of figures after the points in the question.

Dividing

Dividing is also done by changing to fractions, but in a slightly different way. To find 11.421 ÷ 2.7 we say that

$$11.421 \div 2.7 = \frac{11.421}{2.7}$$

$$= \frac{11.421 \times 10}{2.7 \times 10}$$

$$= \frac{114.21}{27}$$

$$= 4.23.$$

This can be simplified as follows:

Example: 11.421 ÷ 2.7

1. Write down the example. — 11.421 ÷ 2.7

2. Multiply both numbers by 10 so that the divisor is a whole number. — 114.21 ÷ 27

3. Carry out the division, and the decimal point in the answer will be in the right place.

```
      4.23
27)114.21
   108
     62
     54
      81
      81
      --
```

Approximations

When multiplying or dividing decimals, a rough check should always be made. Thus, for the examples given:

37.5 × 4.36. Rough approximation, 40 × 4 = 160.
11.421 ÷ 2.7 Rough approximation, 11 ÷ 3 = 4 (roughly).

Exercise 4a

Calculate the following:

(1) 273.2 + 596.3 + 42.91 + 3.72 + 1 034.1

(2) $43.07 + 0.21 + 141.2 + 0.031 + 217.8$
(3) $5.93 + 789.3 + 0.034 + 56$
(4) $4.093 + 27.45 + 1\,030.3 + 4.97$
(5) $61.32 + 4.14 + 273.201 + 44.097$
(6) $72.31 - 59.47$
(7) $98.37 - 49.83$
(8) $43.081 - 27.326$
(9) $54.973 - 35.784$
(10) $101.3 - 87.14$
(11) $73.81 + 97.23 - 42.84 + 27.567 - 83.1$
(12) $21.08 - 43.72 - 21.93 + 78.76 + 103.44$
(13) $907.35 + 146.27 - 1\,096.1 + 142.1 - 23.76$
(14) $54.76 - 93.2 + 48.01 + 76.3 - 20$
(15) $73.2 - 49.26 - 74.81 + 103.2 + 100.57$
(16) 376.3×10
(17) 52.7×100
(18) 842.7×20
(19) 791.3×300
(20) $0.76 \times 4\,000$
(21) $592.6 \div 200$
(22) $841.2 \div 30$
(23) $371.91 \div 700$
(24) $31.76 \div 400$
(25) $72.3 \div 3\,000$
(26) 563.2×54.1
(27) 97.31×4.728
(28) 4.085×127.3
(29) 82.3×0.027
(30) 4.173×3.04
(31) $73.602 \div 1.74$
(32) $1\,705.011 \div 23.1$
(33) $3\,330.13 \div 32.3$
(34) $3\,305.93 \div 106.3$
(35) $3.0744 \div 0.056$

Exercise 4b

(1) If $P = 14.28$ and $Q = 3.4$, find (a) $P + Q$, (b) $P - Q$, (c) $P \times Q$, (d) $P \div Q$.
(2) If $R = 11.5$ and $S = 4.6$, find (a) RS, (b) R^2, (c) $R \div S$, (d) $S \div R$.
(3) A group of boys on holiday walk the following distances between Youth Hostels: Monday 29.7 km, Tuesday 25.4 km, Wednesday 17.2 km, Thursday 23.9 km, Friday 21.5 km.

(a) What is the total distance travelled?

(b) If the sixth hostel is 140.4 km from their starting point, how far must they walk on Saturday?

(c) The boys stay at the same hostel on Saturday night and Sunday night. During Sunday they walk 5.3 km, and the following week they walk back home. What is the total distance walked in the two weeks?

(4) The figure shows six villages A, B, C, D, E and F, visited each day by a postman. Starting at A he drives to each village in turn, taking the route shown by the arrows.

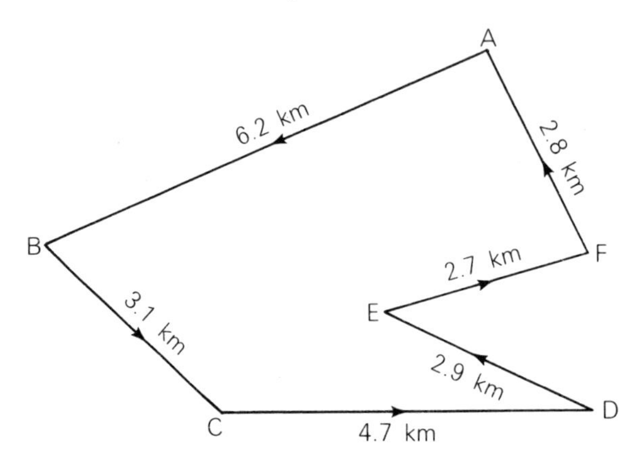

Find (a) the distance travelled each day,

(b) the distance travelled in a six-day week,

(c) the distance travelled in a year in which he works 249 days.

(5) In question 4 the average amount of petrol used is as follows: A to B, 0.66 litres; B to C, 0.29 litres; C to D, 0.45 litres; D to E, 0.32 litres; E to F, 0.30 litres; F to A, 0.32 litres.

Find (a) the amount of petrol used each day,

(b) the amount of petrol used in a month with 26 working days,

(c) how much petrol is left in the tank at the end of the week (assume that the week is started with a full tank holding 35 litres, and that the car is not used for any other journey),

(d) how many complete journeys can be made on a full tank of petrol.

5 Angles and Bearings

Rotation

If you spin the wheel of a bicycle, it rotates about the hub. The amount of rotation can be measured by the number of times the wheel turns. If the wheel just turns far enough to return to its original position, the amount of rotation is one complete turn.

One complete turn is also the amount which the minute hand of a clock turns in one hour. In half an hour it turns through half a turn. In a quarter of an hour it turns through a quarter of a turn. The amount of rotation is called the *angle* of rotation.

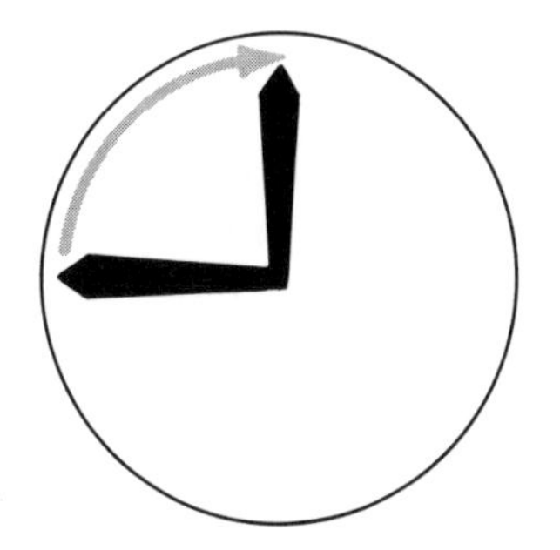

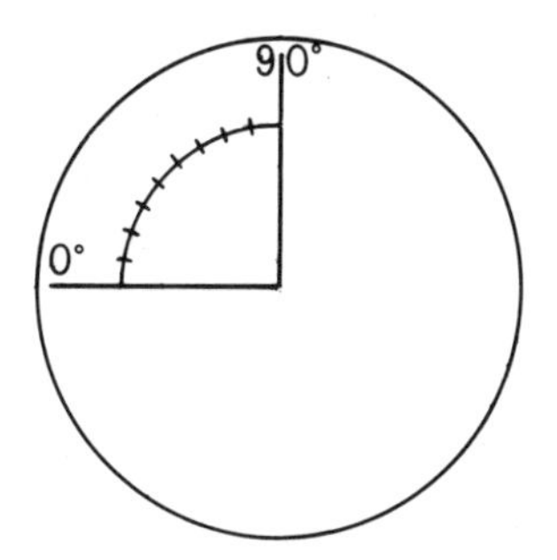

The angle that the minute hand turns through in a quarter of an hour is called a *right angle*. A right angle is divided into ninety equal parts. Each of these parts is called one *degree* (1°). This means that there are $4 \times 90° = 360°$ in one complete turn.

If it is necessary to use a smaller measurement of angle, each degree is divided into sixty equal parts. Each of these parts is called one *minute* (1′).

60 minutes = 1 degree
90 degrees = 1 right angle
4 right angles = 1 complete turn.

Angles are measured with a protractor.

Types of Angle

An angle which is less than one right angle is called an *acute angle.*

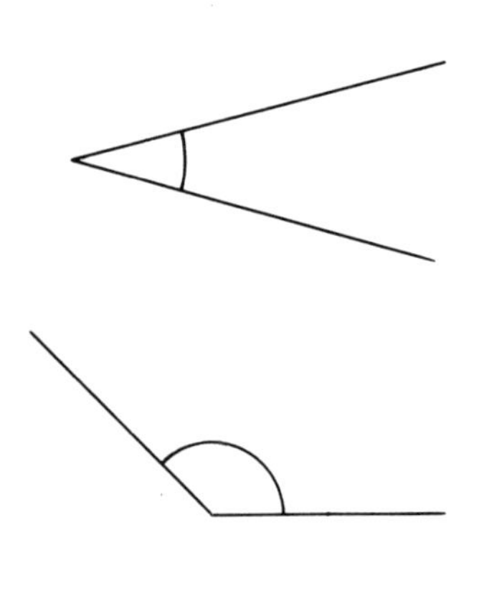

An angle which is more than one right angle but less than two right angles is called an *obtuse angle.*

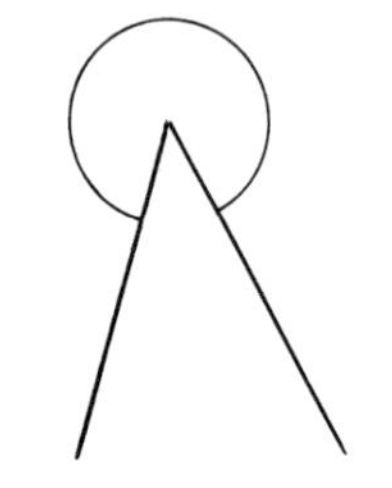

An angle which is more than two right angles but less than one complete turn is called a *reflex angle.*

Points of the Compass

Often we need to know the direction of one place from another. We might say that Edinburgh is roughly 430 km from Bristol. This is not good enough if you are a pilot, and want to fly from Bristol to Edinburgh. You would know how far to go, but you would not know in which direction.

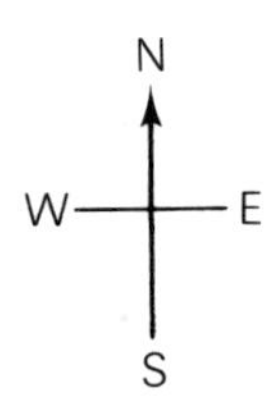

Once you know the distance and the direction, you know exactly where to go. The direction of one place from another is called its *bearing*. The bearing of Edinburgh from Bristol is North. If you are in Edinburgh and want to go to Bristol you must go in the opposite direction. The bearing of Bristol from Edinburgh is South.

In the same way, London is roughly East of Bristol, and Bristol is West of London.

Whole Circle Bearings

Now suppose we are in Bristol and want to know the bearing of Brighton. Imagine standing in Bristol facing North.

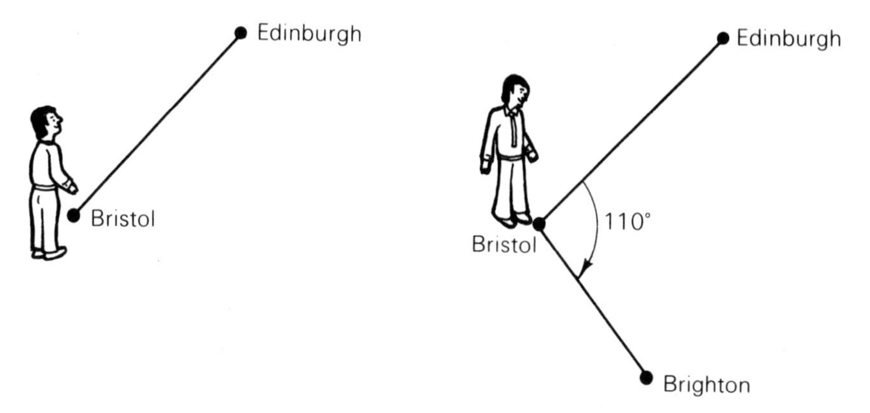

If you turn in a clockwise sense until you face Brighton, the angle you turn through gives the bearing of Brighton. It is 110°.

What is the bearing of Bristol from Brighton? Imagine you are at Brighton facing North. Now turn in a clockwise sense until you face Bristol. What angle would you have turned through?

Bearings are always given as three figure numbers.

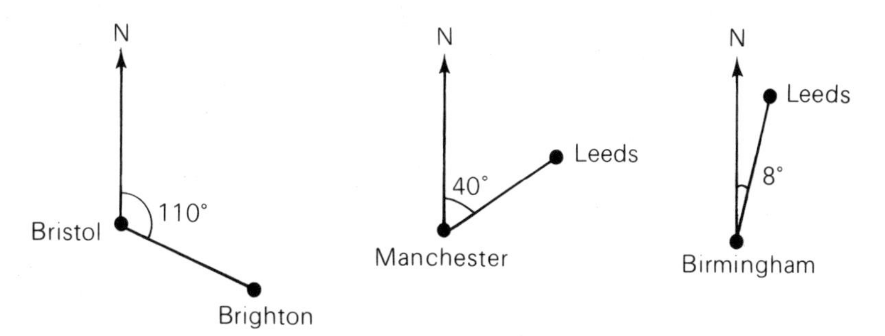

The bearing of Brighton from Bristol is 110°. The bearing of Leeds from Manchester is 040°. The bearing of Leeds from Birmingham is 008°. Bearings given in this way are called *whole circle bearings*.

Quadrant Bearings

Another, and older, method for giving bearings is to measure the angle from North or South. Using this method to find the bearing of Brighton from Bristol, ask yourself whether Brighton is further North or South than Bristol. Brighton is further South.

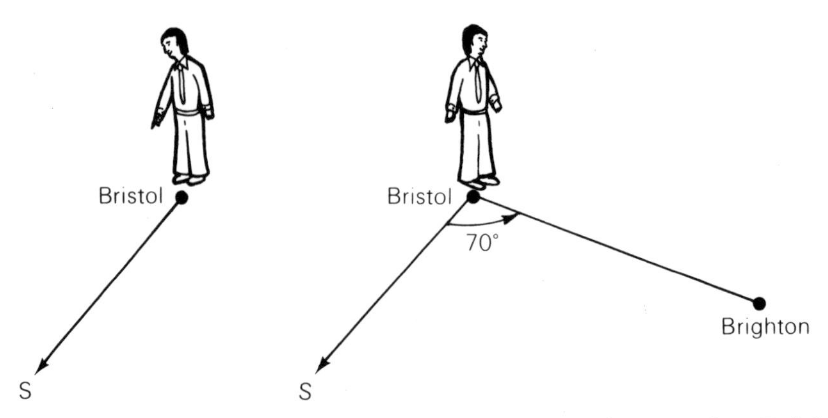

If you start by facing South and then turn so that you face Brighton, the angle you turn through gives the bearing. It is necessary to give the direction you were facing originally (North or South), and then say how much you have turned, and whether you have turned towards East or West. The bearing is given as S 70° E. (South, 70 degrees East.) Bearings given in this way are called *quadrant bearings*. Note that bearings are given from North or South to East or West. You should never give angles starting from East or West.

Exercise 5a

(1) Name the type of angle which is shown in (a) to (f) below.

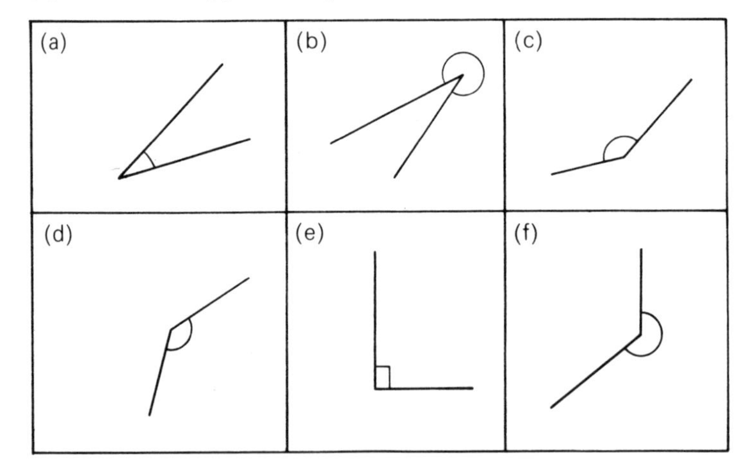

In questions 2–12 look at the clock face on page 27, and write down the angle between the hands at the times which are given.

(2) 3 o'clock
(3) 11 o'clock
(4) 2 o'clock
(5) 1 o'clock
(6) 4 o'clock
(7) 8 o'clock
(8) 7 o'clock
(9) 9 o'clock
(10) 2.30
(11) 1.30
(12) 9.30

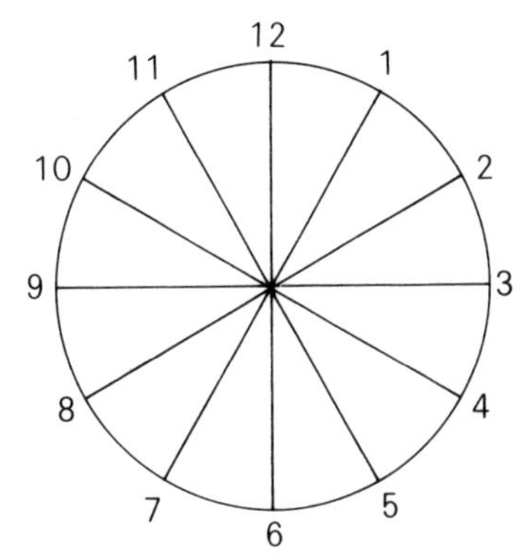

(13) The figure shows some angles with values marked on them. Say whether the values are true or false.

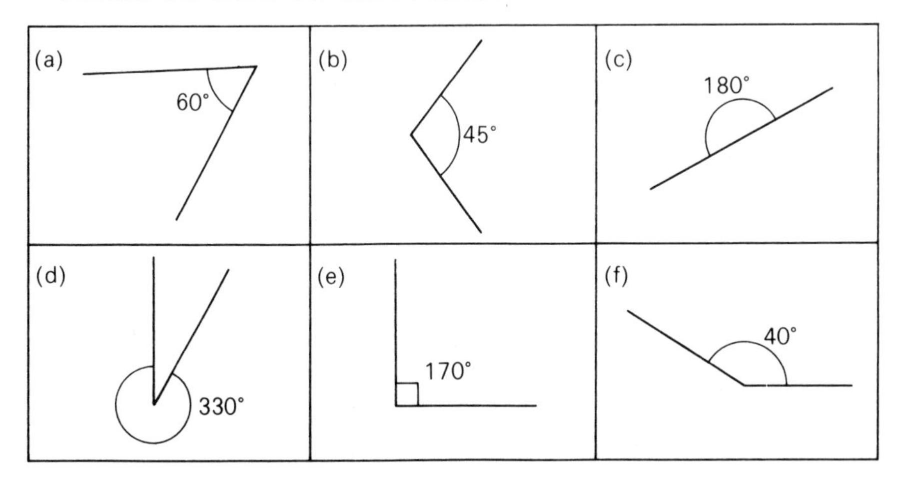

(14) Write down the value of each of the angles marked x in the figure below.

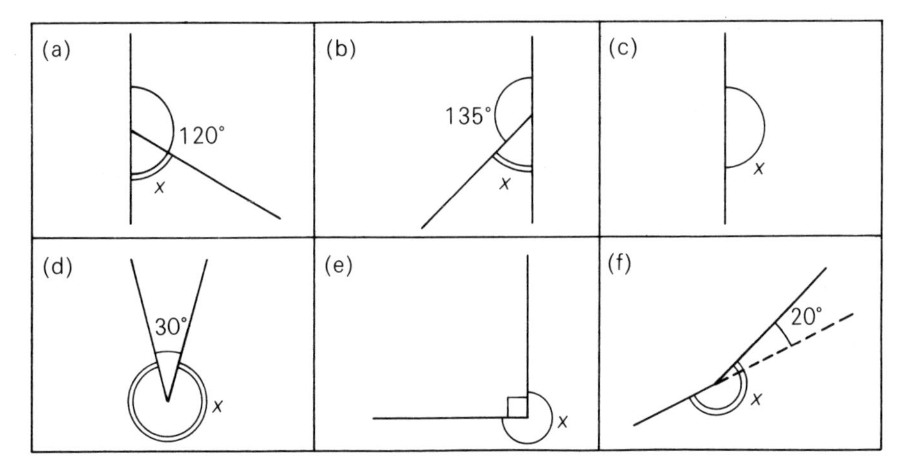

(15) Write down the bearings asked for, using both methods.

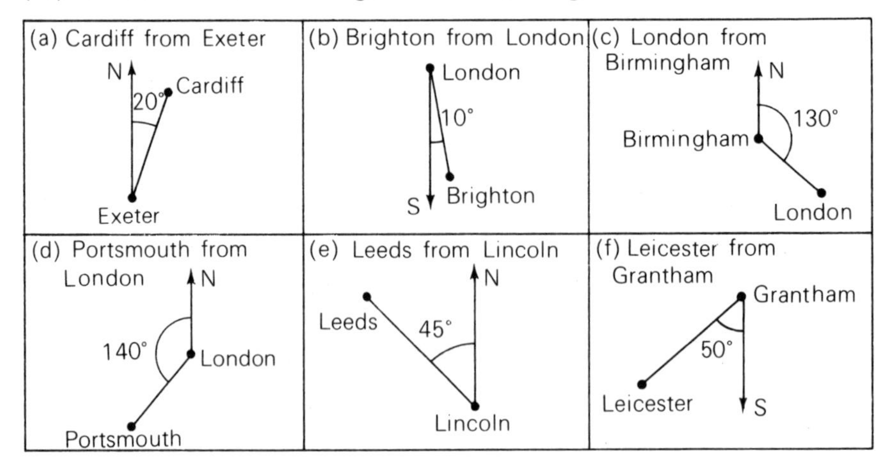

(16) Measure the angles in the figure below, and write down the bearings asked for, using both methods.

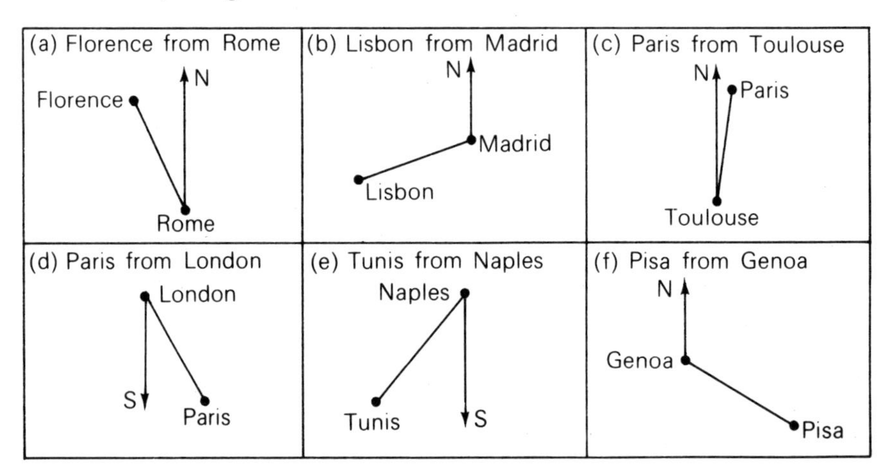

(17) The bearing of one ship from another is 023°. Convert this to a quadrant bearing.
(18) An aircraft flies on a course of N 35° W. Convert this to a whole circle bearing.
(19) Write down the backward bearing for question 17.
(20) Write down the backward bearing for question 18.

Exercise 5b

(1) The figure at the top of page 29 shows the relative positions of Glasgow and Edinburgh.
(a) Use your protractor, and write down the quadrant bearing of Edinburgh from Glasgow.

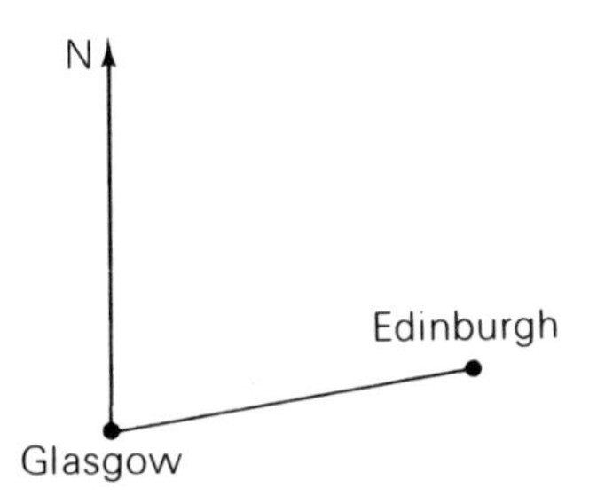

(b) Write down the backward bearing.

(c) A man in Glasgow faces towards Edinburgh. What angle must he turn through in order to face South?

(2) The map shows the relative positions of four towns. Write down the whole circle bearing of

(a) Cheltenham from Gloucester,

(b) Trowbridge from Cheltenham,

(c) Cheltenham from Banbury.

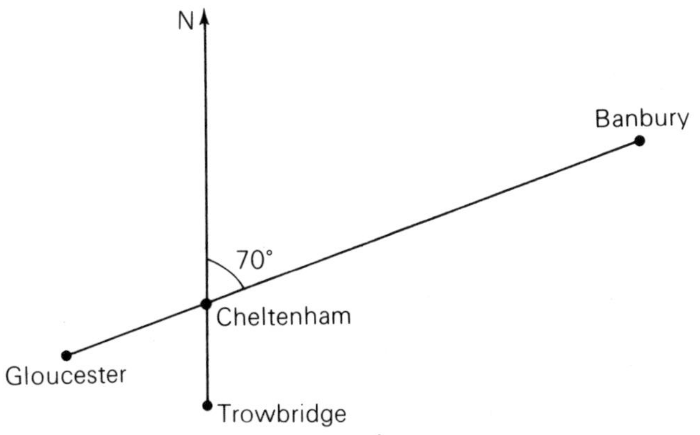

(3) The bearing of Coventry from Oxford is N 12° W. Rugby is North of Oxford.

(a) Draw an accurate diagram showing the bearings of Coventry and Rugby from Oxford.

(b) What is the bearing of Oxford from Coventry?

(c) A small aeroplane leaves Oxford for Coventry, but is blown 20° off course by a wind blowing from the West. On what bearing does the aeroplane fly?

(4) An aeroplane flying on a South-easterly course is blown 8° towards the South by a strong northerly wind.

(a) In which direction does the plane travel?

(b) What is the angle between the direction of the wind and the direction in which the plane travels?

(c) When the wind drops the pilot decides that he must set his course to S 57° E. Through what angle must he turn the plane?

6 Money

British Money

The British system of money is a simple one, in which one hundred pennies equal one pound sterling.

100p = £1.

It follows for example that 1p = £0.01, 23p = £0.23 and so on. One halfpenny is £0.005 and for normal purposes is written as $\frac{1}{2}$p or £0.00$\frac{1}{2}$. It must be realised that the $\frac{1}{2}$p does not have a column of its own. Thus

1st place of decimals
2nd place of decimals
£52.37$\frac{1}{2}$ = £52.375
1st place of decimals
2nd place of decimals
3rd place of decimals

Addition and Subtraction

Apart from the use of the halfpenny this is the same as addition and subtraction of decimals, as the examples show.

£
25.27
3.05
161.81
9.03$\frac{1}{2}$
42.00
£241.16$\frac{1}{2}$

	£
	72.31
−	49.43$\frac{1}{2}$
	£22.87$\frac{1}{2}$

Multiplication and Division

This is usually a matter of multiplying or dividing an amount of money by a whole number. When dealing with large sums of money it is best to write a halfpenny as £0.005 and then work as in chapter 4.

Example: £14.73$\frac{1}{2}$ × 23.

£14.73$\frac{1}{2}$	14 735 × 23	
= £14.735	44 205 294 700 338 905	£14.735 × 23 = £338.905 = £338.90$\frac{1}{2}$.

1. Rewrite the amount changing the halfpenny.

2. Multiply.

3. Replace the decimal point, and change back to halfpennies.

Example: £629.59$\frac{1}{2}$ ÷ 17.

£629.59$\frac{1}{2}$	37.035 17)629.595 51	
= £629.595	119 119 59 51 85 85	£37.035 = £37.03$\frac{1}{2}$.

1. Rewrite the amount changing the halfpenny.

2. Divide.

3. Change back to the halfpenny.

Foreign Money

The list on page 32 shows the money used in some foreign countries. The amount of money which is equal to £1 is called the *exchange rate*. It varies from time to time. Notice that the French and Swiss francs do not have the same value.

Country	Unit	Symbol	Amount equal to £1
USA	Dollar	$	$2.40
France	Franc	Fr	Fr13.50
Italy	Lira	L	L1 555
West Germany	Mark	DM	DM8.50
Switzerland	Franc	f	10.10f

As each of these is a decimal currency, dealing with dollars or francs is no different from dealing with pounds. The Lira however, is so small, being worth less than one tenth of a penny, that whole numbers only need be used.

Example: Change £20.52 to French francs.

£1 = 13.50 francs

∴ £20.52 = 13.50 × 20.52 francs

```
   1 350 ×
   2 052
   -----
   2 700
  67 500
2 700 000
---------
2 770 200
```

£20.52 = 277.02 francs

Example: Change $137.64 to Pounds Sterling ($1 means 1 dollar).

£1 = $2.40

$$\therefore \$137.64 = £\frac{137.64}{2.40}$$

```
     57.35
24)1 376.4
   120
   ---
    176
    168
    ---
      84
      72
      --
      120
      120
      ---
```

$137.64 = £57.35

Exercise 6a

Write the amounts in questions 1–10 in £.

(1) 25p
(2) 87p
(3) 23p
(4) 7p
(5) $51\frac{1}{2}$p
(6) 117p
(7) 193p
(8) $251\frac{1}{2}$p
(9) 703p
(10) $500\frac{1}{2}$p

Write the amounts in questions 11–20 in pence.

(11) £0.63
(12) £0.81
(13) £0.49$\frac{1}{2}$
(14) £0.03
(15) £0.11
(16) £1.26
(17) £8.43
(18) £4.27$\frac{1}{2}$
(19) £11.92$\frac{1}{2}$
(20) £4.03

Calculate the following:

(21) £1.25 + £5.63 + £2.49
(22) £5.07 + £2.40$\frac{1}{2}$ + £13.22
(23) 57p + 39p + 42$\frac{1}{2}$p + 99p
(24) £102.40 + £2 447.33
(25) £58.48 + £97.33$\frac{1}{2}$ + 47p
(26) £7.23 − £3.97
(27) £11.42 − £8.63$\frac{1}{2}$
(28) £10.00 − £3.28$\frac{1}{2}$
(29) £103.25$\frac{1}{2}$ − £87.32$\frac{1}{2}$
(30) £301.10 − £294.27$\frac{1}{2}$
(31) £52.73 × 26
(32) £41.39 × 19
(33) £24.86$\frac{1}{2}$ × 33
(34) £13.03$\frac{1}{2}$ × 17
(35) £15.27$\frac{1}{2}$ × 23
(36) £241.22 ÷ 14
(37) £550.97 ÷ 17
(38) £874.72 ÷ 32
(39) £511.32 ÷ 24
(40) £817.47$\frac{1}{2}$ ÷ 19
(41) $2.27 + $23.05 + $905.23
(42) DM10.76 − DM7.89
(43) L546 − L762 + L1 031
(44) 121.73f × 41
(45) $656.25 ÷ 15
(46) Change £57.50 to Swiss francs.
(47) Change £203.30 to dollars.
(48) Change L8 086 to £.
(49) Change $152.88 to £.
(50) Change DM28.90 to £.

Exercise 6b

(1) In 1975, an American paid the following bills:

Gas: $43.55, $60.15, $122.73, $103.71,
Electricity: $27.63, $9.81, $39.72, $48.03,
Water: $3.25, $3.25, $3.25, $3.25.

(a) What is the total cost of these bills for each quarter of the year?
(b) What is the total fuel bill for the year?
(c) If he saves $50 each month to pay all these bills, how much would he have left over at the end of the year?

(2) A man pays £5.50 each quarter for rent of a telephone. He also pays 2p for each local call.

(a) What is the cost of 112 local calls?

(b) In the first quarter of the year he makes 126 local calls. His long distance calls cost £3.27. What is his phone bill for the quarter?

(c) For the whole year, there are 418 local calls. The cost of long distance calls is £17.38. What is the total phone bill for the year?

(3) A gas boiler uses 516 units of gas in three months.

(a) Find the cost of the gas at $9\frac{1}{2}$p per unit.

(b) Find the cost of the gas at 9p per unit plus a fixed charge of £1.35 per month.

(c) Which method of payment is cheaper for 516 units, (a) or (b)?

(d) Which method of payment is cheaper for 1 000 units?

(4) A British family on holiday in Germany drive a car with a fuel consumption of 13.2 km per litre.

(a) How far did they drive if they used 45 litres?

(b) Find the cost of the petrol in DM at DM0.68 per litre.

(c) What is the cost of the petrol in £? (£1 = DM8.50.)

(5) A tourist in Italy buys the following goods: a handbag which costs L2 826, a pair of shoes which cost L1 948, a skirt which costs L3 100, five postage stamps which cost L260 in all.

(a) What is the total cost of the goods in Lire?

(b) How much change would be expected from L10 000?

(c) What is the equivalent of the change in £?

7 Negative Numbers

Integers

Consider the set of natural numbers

$N = \{1, 2, 3, 4, 5, \ldots\}$.

If we take any pair of natural numbers and subtract one from the other, the result may be another natural number,

$7 - 5 = 2$
$6 - 3 = 3$

or it may be zero,

$6 - 6 = 0$
$4 - 4 = 0$

or it may be a negative number,

$5 - 6 = (-1)$ (negative one)
$7 - 9 = (-2)$ (negative two).

By subtracting any pair of natural numbers, the set of all possible results is a set of numbers called the *integers*.

$Z = \{\ldots (-4), (-3), (-2), (-1), 0, 1, 2, 3, 4, \ldots\}$.

The following example and a few simple rules should help you to use negative numbers correctly.

Adding Negative Numbers

$6 + (-3)$ is the same as $6 - 3$
$6 + (-6)$ is the same as $6 - 6$
$6 + (-9)$ is the same as $6 - 9$.

Examples:

$8 + (-2) = 8 - 2$
$= 6.$

$5 + (-5) = 5 - 5$
$= 0.$

$10 + (-12) = 10 - 12$
$= (-2).$

In each of these examples, the middle step can be left out if you are sure of getting the right answer.

If you are adding two negative numbers, you can add them in the usual way, but keep the negative sign.

Examples:

$(-5) + (-3) = (-8).$

$(-2) + (-1) = (-3).$

Subtracting Negative Numbers

When subtracting negative numbers, there is a simple rule.

Subtracting a negative number is the same as adding a positive number.

Examples:

$3 - (-2) = 3 + 2$
$= 5.$

$7 - (-10) = 7 + 10$
$= 17.$

$(-5) - (-2) = (-5) + 2$
$= (-3).$

$(-6) - (-7) = (-6) + 7$
$= 1.$

As before, the middle step can be left out if you are sure of getting the right answer.

Multiplying and Dividing

When multiplying and dividing two numbers, there is a simple rule for the sign of the result.

If the signs are the same, the result is positive.

If the signs are different, the result is negative.

Examples:
$(-2) \times (-3) = 6$
$3 \times (-4) = (-12)$
$15 \div (-3) = (-5)$
$(-16) \div 4 = (-4)$
$(-30) \div (-6) = 5$

Laws of Integers

The laws of negative numbers are the same as the laws of the positive numbers. If you have forgotten these laws, turn back to chapter 2.

Exercise 7a

Calculate the following:

(1) $7 - 10$
(2) $5 - 6$
(3) $3 - 9$
(4) $5 + (-2)$
(5) $7 + (-5)$
(6) $8 + (-4)$
(7) $6 + (-6)$
(8) $5 + (-5)$
(9) $9 + (-9)$
(10) $5 + (-7)$
(11) $9 + (-10)$
(12) $3 + (-6)$
(13) $(-5) + (-6)$
(14) $(-3) + (-7)$
(15) $(-6) + (-10)$
(16) $(-5) - (-4)$
(17) $(-6) - (-3)$
(18) $(-9) - (-2)$
(19) $(-3) - (-9)$
(20) $(-2) - (-4)$
(21) $(-4) - (-7)$
(22) $(-5) - 8$
(23) $5 + (-3)$
(24) $7 + (-7)$
(25) $8 + (-9)$
(26) $(-4) + (-8)$
(27) $(-4) - (-1)$
(28) $(-7) - (-8)$
(29) $(-5) - (-10)$
(30) $(-6) + (-9)$
(31) $(-4) \times (-3)$
(32) $(-2) \times (-6)$
(33) $(-3) \times (-5)$
(34) $(-2) \times 5$
(35) $(-4) \times 6$
(36) $5 \times (-7)$
(37) $30 \div (-6)$
(38) $20 \div (-4)$
(39) $(-25) \div 5$

(40) $(-24) \div 6$
(41) $(-32) \div (-8)$
(42) $(-15) \div (-3)$
(43) $(-16) \div (-4)$
(44) $2 \times 3 \times (-6)$
(45) $3 \times (-2) \times 4$
(46) $2 \times (-2) \times (-2)$
(47) $(-3) \times (-3) \times (-2)$
(48) $(-4) \times (-1) \times 2$
(49) $\dfrac{(-6) \times (-3)}{9}$
(50) $\dfrac{(-4) \times 10}{(-5)}$.

Exercise 7b

(1) Find the value of:
(a) $7 + (-8)$ (b) $(-5) - (-3)$ (c) $5 \times (-2)$ (d) $(-6) \times (-5)$
(e) $30 \div (-6)$
(2) If $x = (-3)$, find the value of:
(a) $5x + 2$ (b) $4x^2$ (c) $2x^2 + 3x + 1$ (d) $2(x + 1)^2$.
(3) If $a = (-2)$, and $b = 3$, find the value of:
(a) $2a + 3b$ (b) $5b - 7a$ (c) $2a^2 + 3b$ (d) $(a + 2b)^2$.
(4) If $p = (-1)$, $q = (-2)$, find the value of:
(a) pq (b) $q^2 - 5$ (c) $p^2 + 2p$ (d) $p^2 - q^2$.
(5) If $x = (-2)$, $y = (-1)$, $z = 0$, find the value of:
(a) xy (b) yz (c) $x - y + z$ (d) $x^2 + y^2 - z^2$.

Section B
8 Translation, Rotation and Reflection

Translation

Consider a triangle. Any triangle. Move it in any direction you like, as far as you like. But do not turn it.

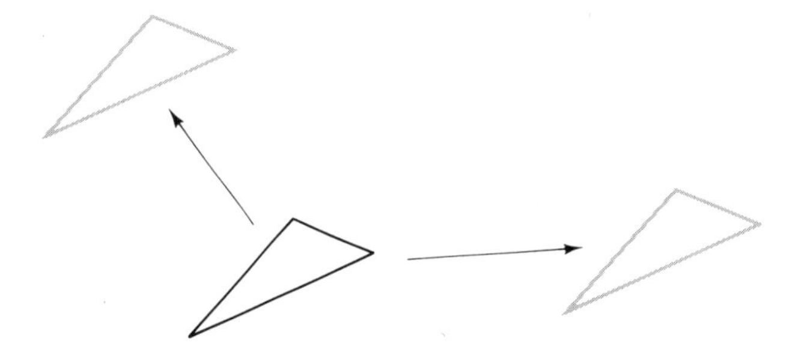

This kind of movement is called a *translation*. The shape of the triangle is still the same. Its size is still the same. Every point on the triangle has moved the same distance in the same direction.

Rotation

Think of the same triangle. This time, do not translate it. Turn it about one of the corners. It can be turned clockwise or anti-clockwise. It can be turned about any point, inside or outside the triangle.

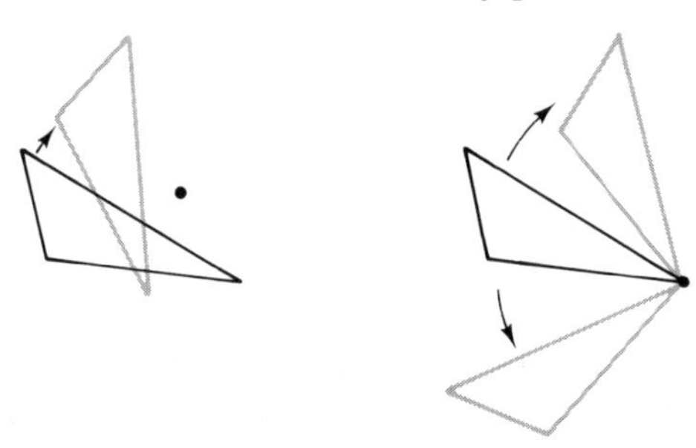

This kind of movement is called *rotation*. The point that the triangle turns about is called the *centre of rotation*.

The shape is still the same. The size is still the same. Every line has been rotated through the same angle. Every point has moved in a circular arc about the centre of rotation.

Reflection

Now hold the triangle up to a mirror. It looks like this.

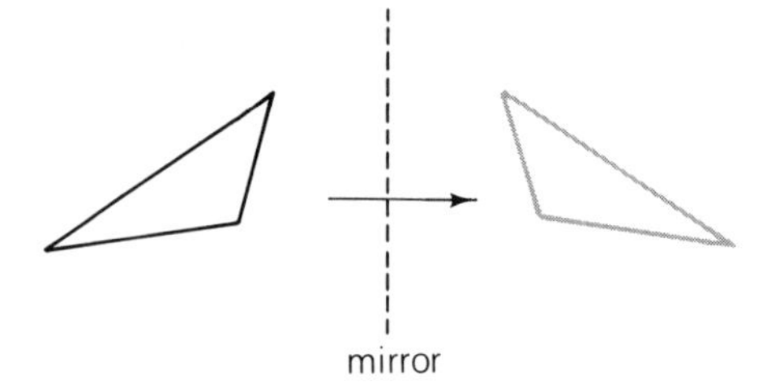

This is a *reflection*. The shape is still the same. The size is still the same. Every point on the triangle has moved in the same direction. It has moved perpendicular to the mirror. Each point on the triangle is the same distance in front of the mirror as its image is behind it.

Line of Symmetry

If an isosceles triangle is reflected along the perpendicular height, each half of the triangle is reflected on to the other half. The mirror line is a *line of symmetry*, or axis of symmetry.

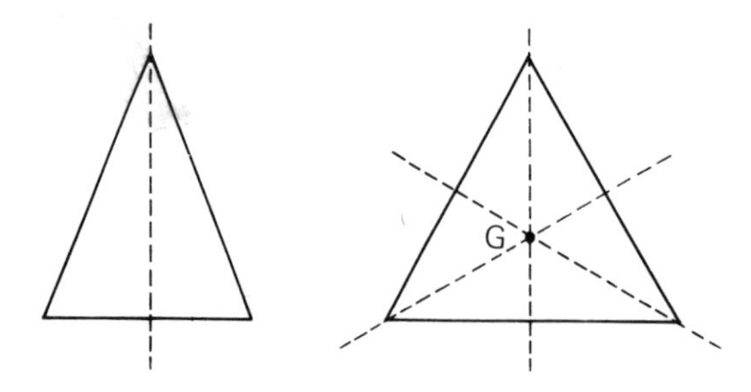

Rotational Symmetry

An equilateral triangle has three lines of symmetry. They all pass through the same point, marked G above. The triangle will map on to itself exactly if it is rotated about G through 120°, or 240°, or 360°. It has *rotational symmetry* of *order* 3.

Point Symmetry

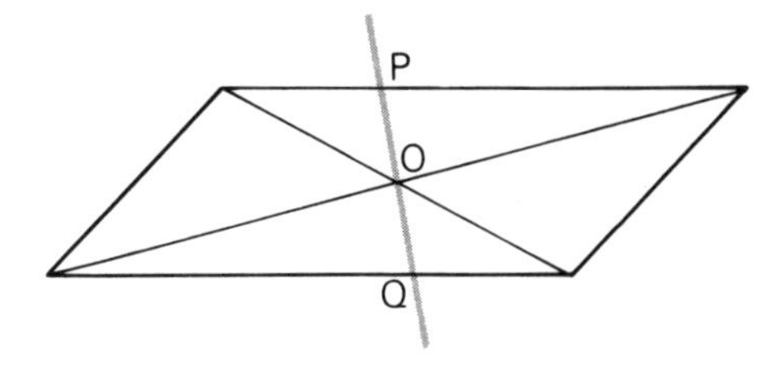

If a parallelogram is rotated about its centre O, through 180° or 360° it maps on to itself exactly. It has rotational symmetry of order 2. This is sometimes referred to as half-turn symmetry, or *point symmetry*. Any line drawn through O cuts the parallelogram at points P and Q such that OP = OQ.

Plane of Symmetry

If a solid object is reflected in a plane mirror, then the mirror is a *plane of symmetry* for the object and its image.

Exercise 8a

(1) Say which kind of movement is shown in each of the following. If there is more than one possible answer, say so.

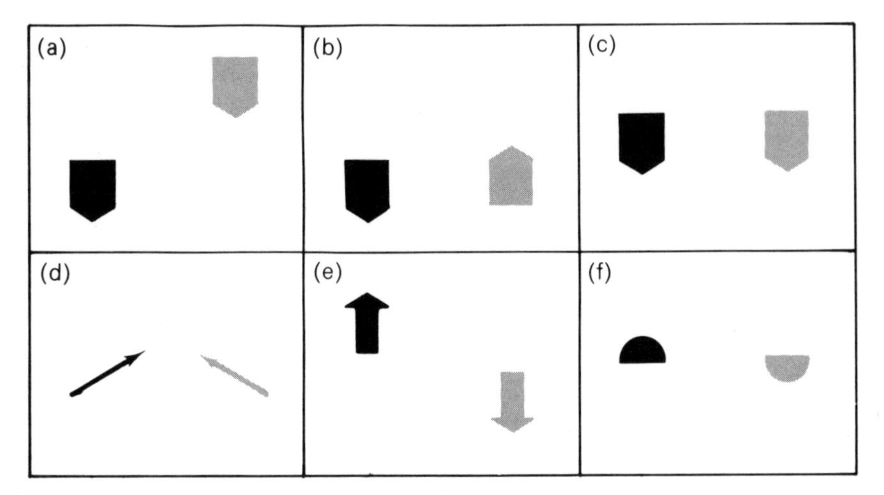

(2) Copy the diagrams and draw lines of symmetry on those figures which have them.

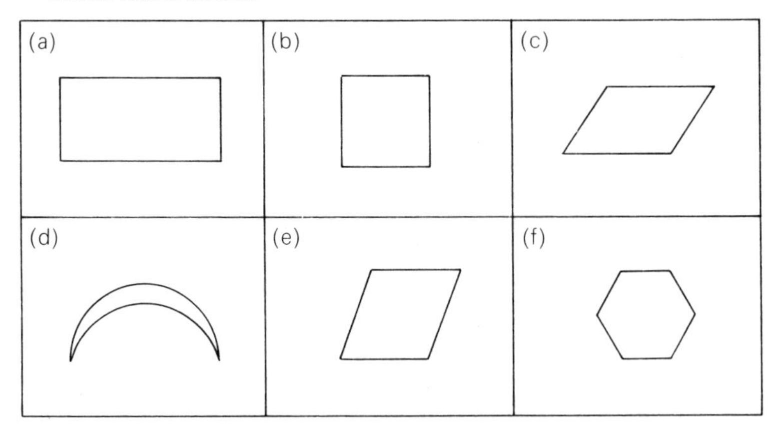

(3) Write down the order of rotational symmetry for each of the following:

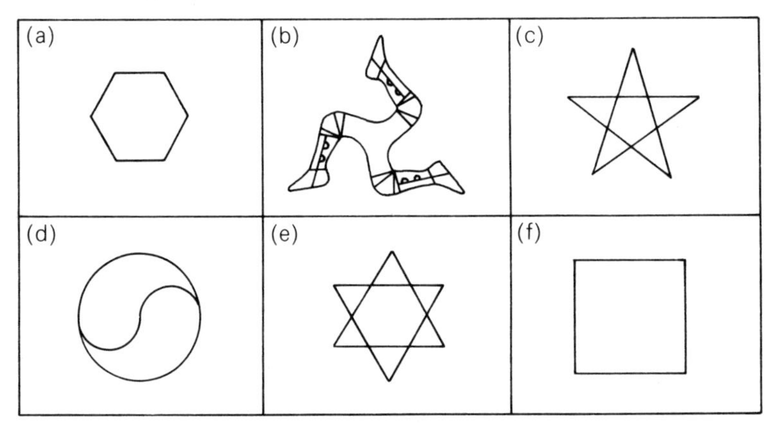

(4) Which of the shapes in question 2 have point symmetry?
(5) Which of the shapes in question 3 have point symmetry?

Exercise 8b

(1) (a) What kind of movement is shown in the figure?
(b) Draw a rectangle and mark on it any lines of symmetry.

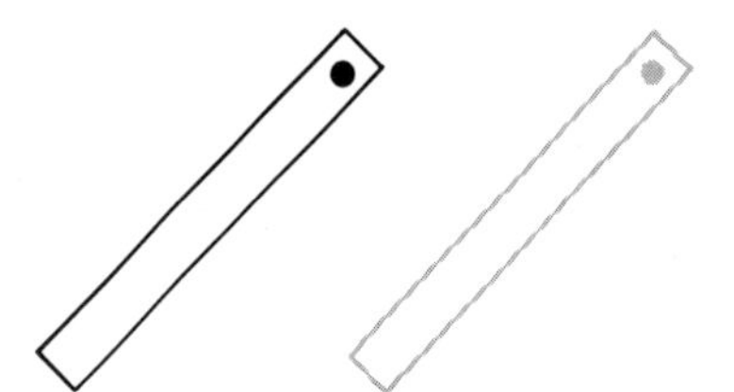

(c) What is the order of rotational symmetry for a rectangle?
(d) Does a rectangle have point symmetry?

(2) (a) What kind of movement is shown in the figure?

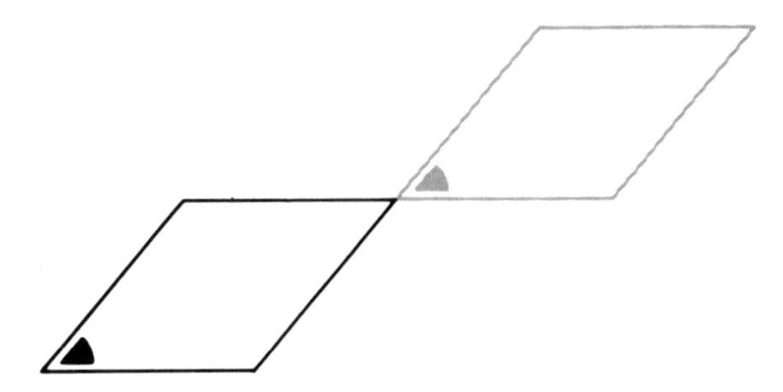

(b) How many lines of symmetry does a parallelogram have?
(c) Does a parallelogram have half-turn symmetry?
(d) Does a parallelogram have point symmetry?

(3) (a) What kind of movement is shown in the figure?

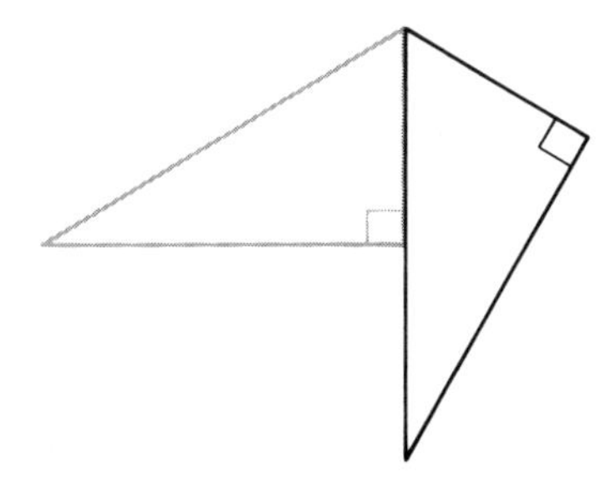

(b) Draw an equilateral triangle and mark the lines of symmetry.
(c) Does an equilateral triangle have point symmetry?
(d) What is the order of rotational symmetry of a regular octagon (8 sides).

(4) (a) The movement shown in the figure could have been made more than one way. Which ways are they?

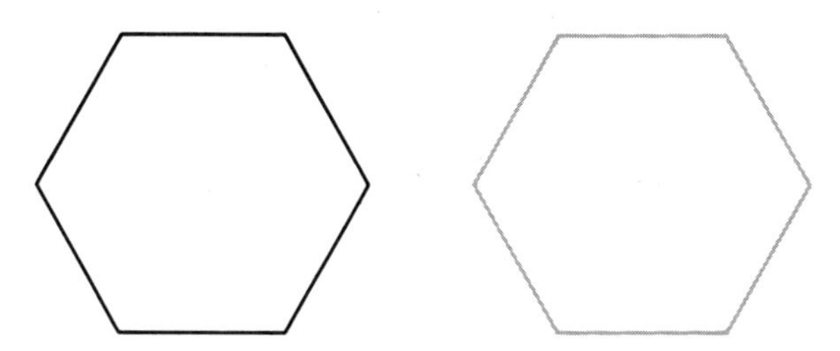

(b) How many lines of symmetry are there to a regular hexagon?
(c) What is the order of rotational symmetry for a regular hexagon?
(d) Does a regular hexagon have point symmetry?

9 Primes, Factors, Multiples

Factors

We know that

$1 \times 72 = 72$
$2 \times 36 = 72$
$3 \times 24 = 72$
$4 \times 18 = 72$
$6 \times 12 = 72$
$8 \times 9 = 72.$

1, 2, 3, 4, 6, 8, 9, 12, 18, 24, 36, 72, are all *factors* of 72.

Prime Numbers

72 has twelve factors. Some numbers have more than this, some have less. Some numbers have only one pair of factors. For example

$1 \times 2 = 2$
$1 \times 3 = 3$
$1 \times 5 = 5$
$1 \times 7 = 7$ and so on.

In each case, the number has only one pair of factors—itself and 1. A number with only one pair of factors is called a *prime number*. Look again at the factors of 72.

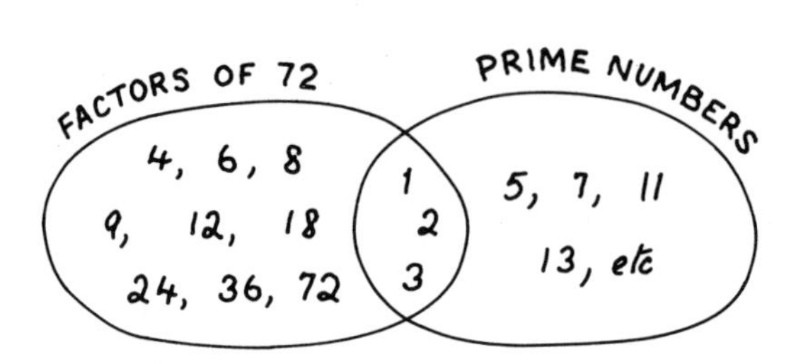

If $F = \{\text{factors of } 72\}$
and $P = \{\text{prime numbers}\}$
then $F \cap P = \{\text{factors of 72 which are prime numbers}\}$.

A factor which is a prime number is called a *prime factor*.

Multiples

If a number is not a prime number, then it has factors. It is a multiple of each of these factors. Look at the number 30.

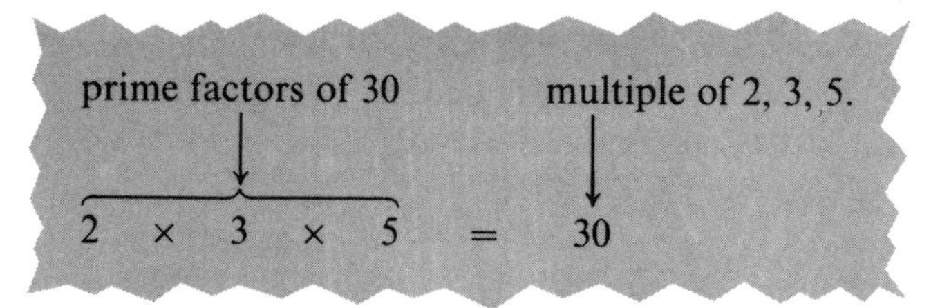

30 is a multiple of 2, 3, and 5. 2, 3, and 5 are all factors of 30. They are all prime factors.

Finding Prime Factors

The prime factors of a number can be found by dividing out. The number can then be written as a product of its prime factors.

Example: Express 924 as a product of its prime factors.

2	924
2	462
3	231
7	77
11	11
	1

$924 = 2^2 \times 3 \times 7 \times 11.$

If the number is even, keep dividing by 2 until an odd number appears. Then try 3, 5, 7, 11, and so on.

Test for Factors

One number is a factor of another if it will divide into it exactly. It is sometimes possible to see if a number is a factor of another without dividing.

If a number is even, then 2 is a factor.
If the digits add up to a multiple of 3, then 3 is a factor.
If 4 is a factor of the last two digits, then it is a factor of the whole number.
If the number ends in a 5 or a 0, then 5 is a factor.
If the digits add up to 9, then 9 is a factor.

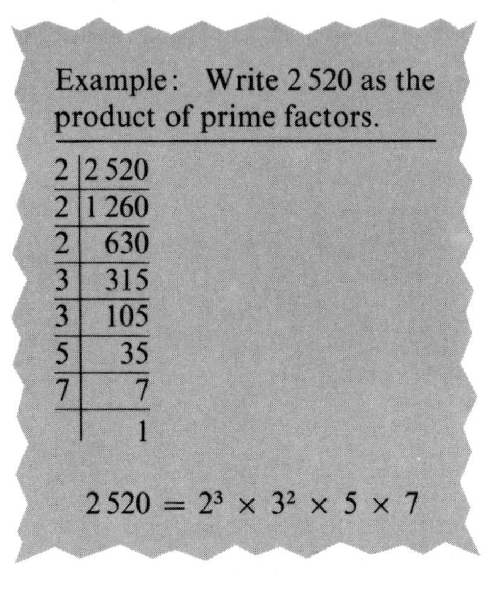

Example: Write 2 520 as the product of prime factors.

2	2 520
2	1 260
2	630
3	315
3	105
5	35
7	7
	1

$2\,520 = 2^3 \times 3^2 \times 5 \times 7$

1. 2 520 is even. Divide by 2.
2. 1 260 is even. Divide by 2.
3. 630 is even. Divide by 2.
4. $3 + 1 + 5 = 9$. Divide by 3.
5. $1 + 0 + 5 = 6$. Divide by 3.
6. 35 ends in 5. Divide by 5.
7. 7 is a prime number.

Square Roots

$8 \times 8 = 64$.

We say that $8^2 = 64$, and that the *square root* of 64 is 8 or $\sqrt{64} = 8$.

If $2 \times 2 \times 2 = 2^3 = 8$.
Then $2^3 \times 2^3 = 2^6 = 64$.
Therefore $\sqrt{2^6} = 2^3$.

If a number is given in index form, its square root can be found by halving the index.

Thus $\sqrt{2^8} = 2^4$
$\sqrt{3^{10}} = 3^5$
and $\sqrt{2^4 \times 5^2} = 2^2 \times 5$.

Example: Find the square root of 540 225.

3	540 225
3	180 075
5	60 025
5	12 005
7	2 401
7	343
7	49
7	7
	1

$$540\,225 = 3^2 \times 5^2 \times 7^4,$$

$$\sqrt{540\,225} = 3 \times 5 \times 7^2 = 735.$$

1. Find the prime factors.

2. Halve each of the indices, and multiply the prime factors.

Factors in Algebra

ab means a multiplied by b. The factors of ab are a and b. In the same way

$$\begin{aligned} abc &= 1 \times abc \\ &= a \times bc \\ &= ab \times c \\ &= ac \times b. \end{aligned}$$

Thus, the factors of abc are a, b, c, ab, ac, bc, as well as abc and 1.

The square root of an algebraic expression can be found, as with numbers, by halving the indices.

$$\sqrt{x^8} = x^4$$

$$\sqrt{p^{10}} = p^5$$

and $\sqrt{x^4y^2} = x^2y$.

Exercise 9a

Write down all the factors of the following:

(1) 36
(2) 32
(3) 40
(4) 56
(5) 50
(6) xyz
(7) a^2b
(8) ab^2
(9) $x(a + b)$
(10) x^2y^2

(11) If $A = \{\text{factors of } 30\}$, and $P = \{\text{prime numbers less than } 10\}$, represent these sets on a Venn diagram, and show the position of each element.
(12) Repeat question 11 for $Q = \{\text{factors of } 56\}$ and $P = \{\text{prime numbers less than } 10\}$.
(13) Repeat question 11 for $L = \{\text{factors of } 210\}$ and $P = \{\text{prime numbers less than } 10\}$.
(14) If $C = \{\text{factors of } 66\}$ and $D = \{\text{factors of } 60\}$, list the elements of $C \cap D$.
(15) Draw a Venn diagram showing {factors of 24} and {factors of 48}.
(16) List the common factors of 24 and 48.
(17) If $R = \{1, 2, 3, 4, 5, 6, 7, 8, 9, 10\}$, which elements of R are multiples of 2?
(18) List the multiples of 5 which are less than 32.
(19) List the multiples of 10 which are less than 55.
(20) List the multiples of 7 which are less than 50.

Say whether the following statements are true or false. Do not divide, but use the tests for factors on page 46.

(21) 10 is a factor of 5 684 720.
(22) 10 is a factor of 64 215.
(23) 2 is a factor of 5 179 643.
(24) 9 is a factor of 442 317 402.
(25) 2 is a factor of 732 516 484.
(26) 4 is a factor of 859 784 232.
(27) 627 514 is a multiple of 3.
(28) 536 427 107 is a multiple of 10.
(29) 376 821 444 is a multiple of 4.
(30) 523 471 113 is a multiple of 6.

Write each of the following as a product of prime factors.

(31) 210
(32) 195
(33) 294
(34) 3 630
(35) 3 150
(36) 120
(37) 2 835
(38) 216
(39) 1 458
(40) 3 750

Find the square root of each of the following:

(41) 900
(42) 1 225
(43) 784
(44) 1 296
(45) 7 056
(46) 2 025
(47) 4 356
(48) 2 704
(49) 2 601
(50) 2 116.

Exercise 9b

(1) (a) Write down all the factors of 60.
(b) Express 60 as a product of prime factors.
(c) Write down any three multiples of 60.
(d) What is the largest factor of 60 which is also a factor of 45?

(2) (a) Write down all the factors of 48.
(b) Express 48 as a product of prime factors.
(c) Which of the following are multiples of 48: 96, 144, 194, 48?
(d) What is the smallest number which is a multiple of 48 and 72?

(3) (a) Write down all the factors of a^2bc.
(b) Write down all the factors of abc^2.
(c) Write down an expression which is a multiple of a^2bc and which is also a multiple of abc^2.

(4) (a) Express 1 764 as a product of prime factors.
(b) Express 3 920 as a product of prime factors.
(c) What is the highest number which is a factor of 1 764 and 3 920?
(d) Use your answer to (a) and write down the square root of 1 764.

(5) (a) Express 194 481 as a product of prime factors.
(b) Find the square root of 194 481.
(c) Find the square root of your answer to (b).

10 Indices

Multiplying with Indices

$x^2 = x \cdot x$
$x^3 = x \cdot x \cdot x$

Multiplying x^2 and x^3 gives

$$x^2 \times x^3 = x \cdot x \times x \cdot x \cdot x$$
$$= x^5.$$

Thus, to multiply, add the indices. If there is more than one letter, they must be multiplied separately.

Example: $4x^3y^2 \times 5x^4y^3$

1. Write down the question.

$4x^3y^2 \times 5x^4y^3 = ■$

2. Multiply the numbers and write down the answer.

$4x^3y^2 \times 5x^4y^3 = 20■$

3. Multiply the x's in your head, and write down the answer.

$4x^3y^2 \times 5x^4y^3 = 20x^7■$

4. Multiply the y's in your head and write down the answer.

$4x^3y^2 \times 5x^4y^3 = 20x^7y^5.$

Division with Indices

$x^7 = x \cdot x \cdot x \cdot x \cdot x \cdot x \cdot x$
$x^5 = x \cdot x \cdot x \cdot x \cdot x$

Dividing x^7 by x^5 gives

$$\frac{x^7}{x^5} = \frac{x \cdot x \cdot x \cdot x \cdot x \cdot x \cdot x}{x \cdot x \cdot x \cdot x \cdot x}$$

$$= \frac{\not{x} \cdot \not{x} \cdot \not{x} \cdot \not{x} \cdot \not{x} \cdot x \cdot x}{\not{x} \cdot \not{x} \cdot \not{x} \cdot \not{x} \cdot \not{x}}$$

$$= x \cdot x$$

$$= x^2$$

Thus, to divide, subtract the indices. If there is more than one letter, these must be divided separately.

Example: $12x^4y^5 \div 3x^3y^2$

1. Write down the question. $12x^4y^5 \div 3x^3y^2 = ■$

2. Divide the numbers and write down the answer. $12x^4y^5 \div 3x^3y^2 = 4■$

3. Divide the x's in your head, and write down the answer. $12x^4y^5 \div 3x^3y^2 = 4x■$

4. Divide the y's in your head, and write down the answer. $12x^4y^5 \div 3x^3y^2 = 4xy^3$.

Squares and Square Roots

In order to square an algebraic expression, square the coefficient, and double the indices of the letters.

Example: $(4x^3y^4)^2$

1. Square the coefficient and write down the answer. $(4x^3y^4)^2 = 16■$

2. Double the indices and write down the answer. $(4x^3y^4)^2 = 16x^6y^8$.

In order to find the square root of an expression, find the square root of the coefficient and halve the indices.

1. Find the square root of the coefficient and write down the answer.

2. Halve the indices and write down the answer.

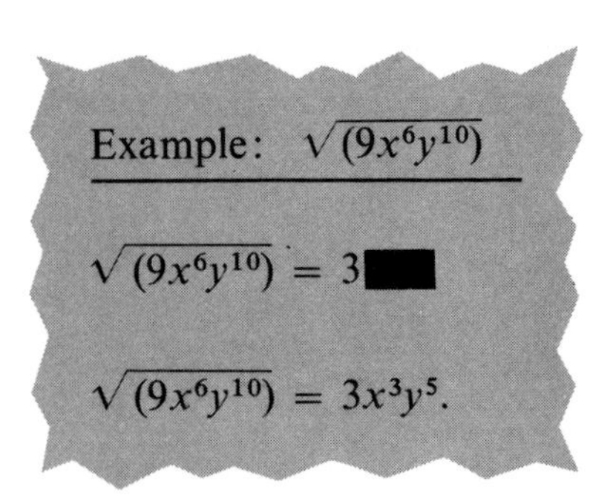

Exercise 10a

Simplify the following:

(1) $x^3 \times x^4$
(2) $x^5 \times x^2$
(3) $y^{10} \times y^3$
(4) $p \times p^6$
(5) $a^7 \times a^4$
(6) $x^2 \times x^3 \times x^4$
(7) $a^4 \times a \times a^6$
(8) $2x^4 \times 5x^3$
(9) $6x^2 \times 3x^5$
(10) $2a^2 \times 4a^3$
(11) $2x^2y \times 3xy^2$
(12) $4x^2y^2 \times 2x^2y^3$
(13) $3a^2b^2 \times 2a^5b^4$
(14) $6x^{10}y \times 3x^2y^5$
(15) $4x^7y^2 \times 2xy^3z^2$
(16) $x^9 \div x^7$
(17) $a^5 \div a^3$
(18) $b^6 \div b^6$
(19) $4x^6 \div 2x^2$
(20) $20y^9 \div 4y^3$
(21) $4x^2y^5 \div 2xy^3$
(22) $15x^4y^2 \div 3xy^2$
(23) $36x^8y^4 \div 9x^7y$
(24) $35a^7b^{12}c^3 \div 5a^5b^4c^2$
(25) $14p^9q^6r^2 \div 2pq$
(26) $(x^3)^2$
(27) $(a^2b^3)^2$
(28) $(2p^2q)^2$
(29) $(3x^2y^3)^2$
(30) $(5a^3bc^2)^2$
(31) $(3x^2y^5z)^2$
(32) $(9a^5b^5c^2)^2$
(33) $\sqrt{(9a^2b^4)}$
(34) $\sqrt{(4x^4y^6)}$
(35) $\sqrt{(16p^2q^8r^2)}$
(36) $\sqrt{(36l^4m^{10})}$
(37) $\sqrt{(64x^2y^8)}$
(38) $\sqrt{(49a^6b^6)}$
(39) $\sqrt{(25a^{12}b^6)}$
(40) $\sqrt{(81p^4q^6)}$.

Exercise 10b

(1) If $P = 18x^2y^2z^4$, $Q = 3xy^2z^2$, find: (a) PQ (b) $P \div Q$ (c) P^2 (d) Q^2.

(2) If $P = 36a^6b^8c^2$, $Q = 3ab^3c$, find: (a) PQ (b) $P \div Q$ (c) Q^2 (d) $\sqrt{P}$.

(3) If $A = 2p^2q^2r$, $B = 15pr^2$, $C = 6qr$, find: (a) $A \times B \times C$ (b) $\dfrac{A \times B}{C}$ (c) A^2 (d) $3A^2 \div C$.

(4) If $P = 18x^3y^4z^5$, $Q = 2xy^2z^3$, find: (a) $P \times Q$ (b) $(PQ)^2$ (c) $\sqrt{PQ}$ (d) $P \div Q$.

(5) If $X = 8ab^2c$, $Y = 2ac$, find: (a) X^2 (b) Y^2 (c) $X^2 \times Y^2$ (d) $(XY)^2$ (e) $\sqrt{\dfrac{X}{Y}}$

11 Rates and Taxes

Central and Local Government

The country as a whole is governed by Parliament at Westminster. This is called the *central government*. The central government provides many services for the whole country, such as defence. In order to pay for these things money is obtained by such means as income tax and the value added tax (VAT).

Local services such as housing and education are provided by local councils. The council for your district is your *local government*. In order to pay for local government services, the local authority collects money from the people who live in its area.

Payments made to the local government are called *rates*. Payments made to the central government are called *taxes*.

Rates

Every householder pays rates. Rates are also paid for shops and factories. The amount to be paid in rates depends on the kind of property.

When a local authority wants to decide the amount to be paid in rates by each householder, every property in its area is inspected, and given a value. This is called the *rateable value* (RV), of the property.

If the rateable value of your house is £100 and the rateable value of another house is £50, it does not mean that this is what the houses are worth, or that this is what you each pay in rates. It simply means that you will have to pay twice as much in rates as the owner of the other house.

When the local authority has decided on the rateable value of every property in its area, it adds all of these rateable values together, to get the total rateable value for the area. It must then work out how much money will be needed for the year to pay for the services which it provides. If the amount of money which is needed is 67% of the total rateable value for the area then each person pays 67% of his rateable value to the authority.

For a house with RV £100, the rates would then be £67 each year. For a house with rateable value £50, the rates would be half of this (£33.50). For each pound of rateable value, the householder would pay 67p. He would pay 67p in the £.

Amount paid in rates = rateable value × rate in the £.

Example: Find the rates paid by a householder if his RV is £66 and the rate is 53p in the £.

Amount paid is £66 × 0.53 = £34.98.

The authority can work out its total income in rates in the same way.

Total income in rates = total rateable value × rate in the £.

Example: The total rateable value of a town is £750 000. What is its annual income when the rate is 49p in the £?

Total income is £750 000 × 0.49 = £367 500.

Suppose the total rateable value of a town is £930 000, and the rate is 1p in the £. The income of the town would then be £930 000 × 0.01 = £9 300. This is known as the *product of a penny rate*. It shows how much extra income the authority could get by putting the rates up by 1p in the £.

Income Tax

The total amount which a man earns is called his *gross income*. The amount of income tax which he pays depends on his income, but tax is not paid on the gross income. Before tax is calculated some deductions called *allowances* are made from the gross income and tax is then paid on the remainder.

Gross income − allowances = taxable income.

The deductions which are made are fixed in April each year by the central government. They vary from year to year and the following are the kind of amounts which are allowed.

For a single person	£675
For a married man	£955
For each child under 11 years	£240
For each child over 11 and under 16 years	£275

There are also other allowances, and these vary too from time to time.

Example: A man is married and has 2 children, both under 11 years of age. Calculate his taxable income if his gross income is £2 100.

Married man's allowance	£955
Allowance for children	
£240 × 2	£480
Total	£1 435

Taxable income = gross income − allowances
= £2 100 − £1 435
= £665

Once the allowances have been deducted, the tax which is to be paid is calculated as a percentage of the taxable income, in the same way that rates are calculated as a percentage of the rateable value of the house.

Example: Calculate the income tax paid by a man in a year in which his taxable income was £1 500 and the rate of tax was 32p in the £.

Tax paid is £1 500 × 0.32 = £480

Exercise 11a

In questions 1–5 find the rates paid.

(1) RV £100, @ 50p in the £.
(2) RV £96, @ 75p in the £.
(3) RV £95, @ 40p in the £.
(4) RV £360, @ 55p in the £.
(5) RV £171, @ 67p in the £.

In questions 6–10, find the rateable value if the amount paid is

(6) £81.32, @ 76p in the £.
(7) £43.35, @ 51p in the £.
(8) £44.64, @ 48p in the £.
(9) £64.99, @ 97p in the £.
(10) £121.41, @ 57p in the £.

In questions 11–15, find the total income of the authority if the total RV is

(11) £763 000, @ 63p in the £.
(12) £687 500, @ 74p in the £.
(13) £958 250, @ 84p in the £.
(14) £1 532 400, @ 49p in the £.
(15) £1 014 500, @ 67p in the £.

In questions 16–20 calculate the taxable income.

(16) Gross income £1 700, allowances £460.
(17) Gross income £7 500, allowances £675.
(18) Gross income £2 100, allowances £960.
(19) Gross income £3 475, allowances £1 427.
(20) Gross income £1 952, allowances £478.

In questions 21–25, use the table of allowances given on page 55 to calculate the total deductions allowed for tax purposes.

(21) A single man with £75 allowed for a housekeeper.
(22) A married man with three children under 11 years.
(23) A married man with no children, who is allowed £75 for dependent relatives.
(24) A married man with one 14 year old child, and who is allowed £75 for life insurance.
(25) A married man with 4 children under 11 years and one 12 year old child.

In questions 26–30, calculate the tax paid if the taxable income is

(26) £3 177, @ 30p in the £.
(27) £1 975, @ 33p in the £.
(28) £5 044, @ 32p in the £.
(29) £4 772, @ 31p in the £.
(30) £1 725, @ 34p in the £.

Exercise 11b

Use the table of tax allowances on page 55 where necessary.

(1) Mr Richardson lives in a house of RV £324 in an area where the total rateable value is £525 000. If the general rate is 65p in the £. Calculate
(a) the general rate paid by Mr Richardson each year

(b) the water rate, paid by Mr Richardson if it is 5% of the general rate

(c) the total income from the general rate in the whole area

(d) the product of a penny rate.

(2) The total rateable value of the property in a rating area is £760 000 and the general rate is assessed at 73p in the £. Calculate

(a) the product of a penny rate

(b) the total income of the authority

(c) the cost to the authority of maintaining police services at the rate of 14p in the £ of the total rateable value.

(3) A married man with one 8 year old child has a gross income of £4 410 and is allowed £65 to set against income tax for life insurance premiums. Calculate

(a) the total allowances to be set against tax

(b) the taxable income

(c) the tax paid at 35p in the £

(d) the percentage of his gross income which he pays in tax.

(4) A householder lives in a house with RV £573. His gross income is £3 500 and his allowances to be set against his gross income for tax purposes amount to £1 450. Calculate

(a) his taxable income

(b) the tax paid at 33p in the £

(c) the rates paid at 76p in the £

(d) the total amount paid in taxes and rates.

(5) The rateable value of a property is £97 and the rate is 48p in the £ and the householder has a taxable income of £473. Calculate

(a) the general rate paid

(b) the water rate, which is $12\frac{1}{2}\%$ of the general rate

(c) the amount paid in income tax at 35p in the £

(d) the total amount paid in rates and taxes.

12 Brackets

Single Bracket

When multiplying a bracket by a number, remember that every term inside the bracket must be multiplied by the number. If you have forgotten about this, go back to the laws of numbers in chapter 2.

Example: $4(3x + 2)$

1. Multiply the $3x$ by the 4 and write down the answer.

$4(3x + 2) = 12x$ ■

2. Multiply the $+2$ by the 4 and write down the answer.

$4(3x + 2) = 12x + 8$

Example: $-3(2x - 1)$

1. Multiply the $2x$ by the -3 and write down the answer.

$-3(2x - 1) = -6x$ ■

2. Multiply the -1 by the -3 and write down the answer.

$-3(2x - 1) = -6x + 3$

Simplification

When simplifying an expression, only like terms can be added together. Numbers cannot be added to x, and x cannot be added to x^2.

Example: Simplify the expression $2(1 + x^2) - 3x(2 - x)$.

$$2(1 + x^2) - 3x(2 - x) = 2 + 2x^2 - 6x + 3x^2$$
$$= 5x^2 - 6x + 2.$$

In this example the answer has been arranged in *descending powers* of x. This means that the answer starts with the highest power of x, which is x^2. This is followed by the next highest power, and so on. The expression finishes with the constant.

Multiplying Brackets

When you multiply two brackets together, everything in the first bracket must be multiplied by everything in the second bracket.

$$(a + b)(c + d) = ac + ad + bc + bd.$$

Example: $(2x + 3)(3x - 2)$

1. Multiply $2x$ by $3x$ (giving $6x^2$), and $2x$ by -2 (giving $-4x$).

$$(2x + 3)(3x - 2) = 6x^2 - 4x$$

2. Multiply 3 by $3x$ (giving $9x$), and 3 by -2 (giving -6).

$$(2x + 3)(3x - 2) = 6x^2 - 4x + 9x - 6$$

3. Collect like terms.

$$(2x + 3)(3x - 2) = 6x^2 - 4x + 9x - 6 = 6x^2 + 5x - 6$$

Exercise 12a

In questions 1–15 multiply, and so remove the brackets.

(1) $3(x + 5)$
(2) $7(x - 2)$
(3) $3(x + 4)$
(4) $4(2x - 3)$
(5) $3(2x + 7)$
(6) $2(5x - 3)$
(7) $2x(x + 1)$
(8) $3x(x - 2)$
(9) $x(4x + 7)$
(10) $3x(2x + 4)$
(11) $-2x(x + 6)$
(12) $-3x(3x - 7)$
(13) $4x(x + y)$
(14) $-x(3 + 2y)$
(15) $-(a - b)$

Simplify the expressions in questions 16–30.

(16) $2(x + 1) + 5(x + 3)$
(17) $3(x + 5) + 2(2x + 1)$
(18) $5(2x + 1) + 2(3x - 1)$
(19) $7(x + 3) + 3(2x - 3)$
(20) $4x(2x - 3) + 3(2x + 1)$
(21) $6x(x - 1) + 2(3x + 4)$
(22) $2x(2x - 1) + 3(5x - 1)$
(23) $4x(x - 1) + 2x(3x - 2)$
(24) $5(a + 3) - 2(a - 6)$
(25) $4(p + 3) - 3(2p - 4)$

(26) $6x(x + 1) - 3(x - 1)$
(27) $2x(x - 1) - 4(x + 3)$
(28) $2x(x - 1) - 3x(x + 1)$
(29) $5x(x + 2) + x(3x + 7)$
(30) $4x(x - 3) - (x + 6)$

In questions 31–45 multiply the brackets.

(31) $(x + 1)(x + 3)$
(32) $(2x + 2)(x + 4)$
(33) $(x + 5)(3x + 3)$
(34) $(3x + 6)(2x + 1)$
(35) $(4x + 3)(x - 2)$
(36) $(4x + 6)(x - 2)$
(37) $(x + 2)(2x - 3)$
(38) $(3x + 5)(2x - 1)$
(39) $(x - 1)(x + 1)$
(40) $(3x - 2)(x + 1)$
(41) $(2x - 5)(4x + 4)$
(42) $(2x - 3)(4x + 1)$
(43) $(3x - 3)(x - 6)$
(44) $(5x - 2)(x - 3)$
(45) $(x - 7)(3x - 4)$

Exercise 12b

(1) Simplify
(a) $2(3x + 2) + 3(x + 1)$
(b) $3(2x - 1) - 2(2x - 3)$.
Multiply
(c) $(x + 1)(x + 3)$
(d) $(2x + 1)(3x - 4)$.

(2) If $P = (5x + 3)$ and $Q = (2x - 7)$, find
(a) $-3P$
(b) $P + Q$
(c) $2P - Q$
(d) PQ.

(3) If $A = (2x - 3)$ and $B = (3x + 1)$, find
(a) $-2A$
(b) $B - A$
(c) AB
(d) B^2.

(4) If $L = (3x + 4)$ and $M = (2x - 3)$, find
(a) $L + M$
(b) $L - 2M$
(c) LM
(d) L^2.

(5) If $P = (2x - 1)$ and $Q = (2x + 1)$, find
(a) $P + Q$
(b) $3P - Q$
(c) PQ
(d) P^2.

13 Angles and Parallels

Angles at a Point

If two straight lines cross, they form *vertically opposite angles.*

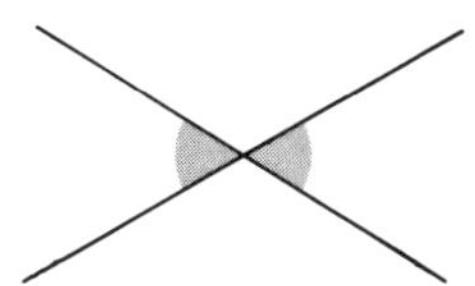

Vertically opposite angles are equal.

If one straight line meets another as shown, the two angles are *adjacent angles* on a straight line.

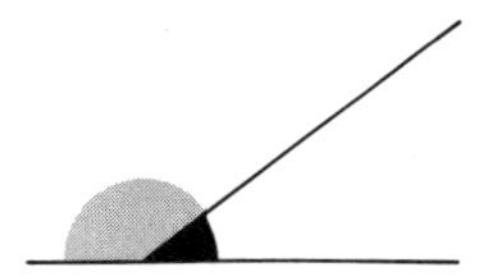

Adjacent angles on a straight line add up to 180°.

Parallels and Transversals

An angle is a measure of rotation. If one line can be mapped to another line by translation, then the two lines are *parallel.* Parallel lines are shown by marking them with arrows. A line which cuts a set of parallel lines is called a *transversal.* In cutting parallel lines, a transversal forms pairs of angles with known properties.

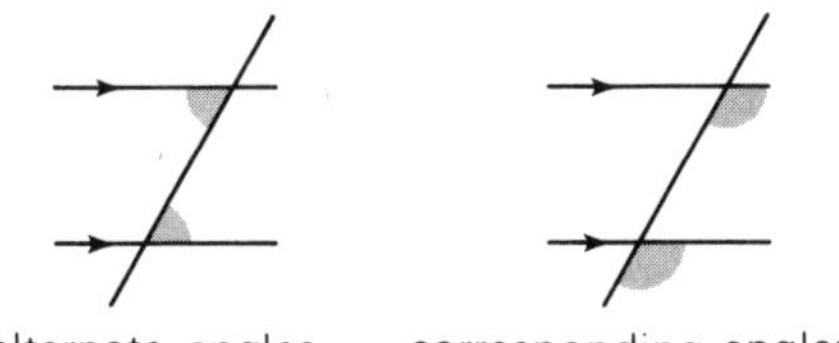

alternate angles corresponding angles interior angles

Alternate angles are equal. Corresponding angles are equal. Interior angles add up to 180°. In the figure on page 61, angles which look the same *are* the same. Angles which look different add up to 180°.

Angles of a Triangle

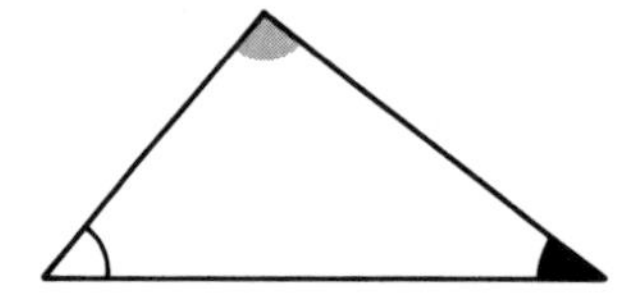

No matter what shape a triangle is, its angles add up to 180°. The largest angle of a triangle is opposite the largest side. The smallest angle is opposite the smallest side.

Types of Triangle

A triangle with all angles less than 90° is an *acute-angled triangle*.
A triangle with a right angle is a *right-angled triangle*.
A triangle with an obtuse angle is an *obtuse-angled triangle*.
A triangle with all sides different is a *scalene triangle*.
A triangle with two equal sides is an *isosceles triangle*.
A triangle with three equal sides is an *equilateral triangle*.
An isosceles triangle has two angles equal. An equilateral triangle has all three angles equal.

Exercise 13a

(1) Write down the value of the angle marked x in each of the following:

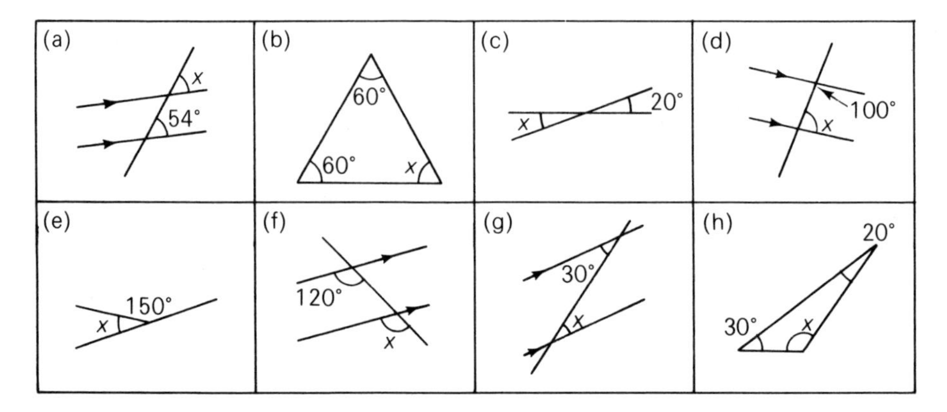

(2) Two angles of a triangle are 30° and 40°. What is the third angle?
(3) Two angles of a triangle are 70° and 100°. What is the third angle?
(4) A triangle has all three angles equal. What is the size of each angle?
(5) One angle of a triangle is 40°. The other two angles are equal. What size are they?
(6) Two equal angles of a triangle add up to the third angle. What is the size of the largest angle?
(7) The largest angle of an isosceles triangle is equal to the sum of the other two. What is the size of the largest angle?
(8) One angle of a triangle is twice as large as the other two put together. What is the size of the largest angle?
(9) Two angles of a triangle add up to 90°. What is the size of the third angle?
(10) In $\triangle ABC$, $\angle A = \angle B + 30°$, and $\angle B = \angle C + 30°$. What is the size of $\angle A$?

Exercise 13b

(1) A triangle has $\angle BAC = 30°$ and $\angle ABC = 40°$,
(a) calculate $\angle ACB$,
(b) say whether the triangle is acute-angled, right-angled, or obtuse-angled,
(c) say whether the triangle is scalene, isosceles, or equilateral,
(d) which is the longest side?
(2) Look at the diagram, and write down the value of:

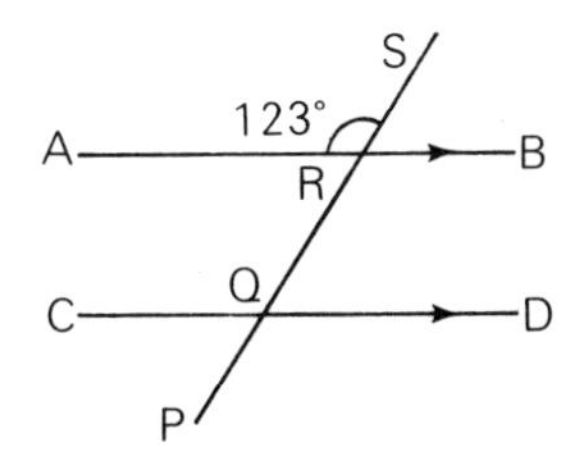

(a) $\angle PRB$ (b) $\angle CQS$ (c) $\angle SRB$ (d) $\angle PQD$.
(3) Look at the diagram and write down the value of:

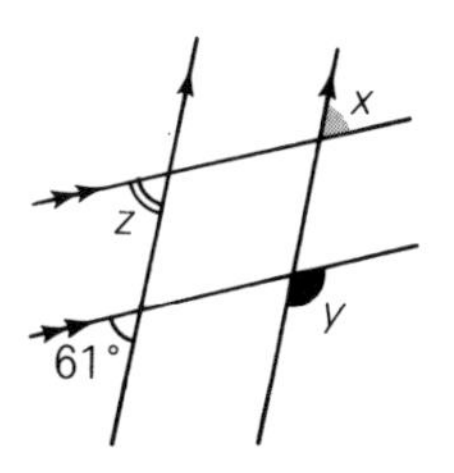

(a) $\angle x$ (b) $\angle y$ (c) $\angle z$.

(4) Look at the diagram and write down the value of:

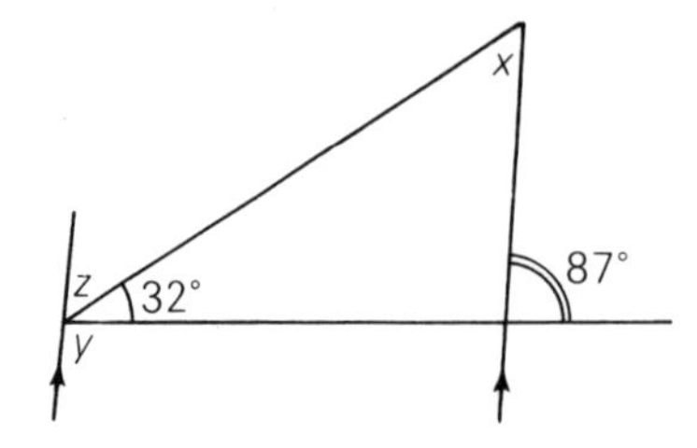

(a) $\angle x$ (b) $\angle y$ (c) $\angle z$.

(5) In the diagram,

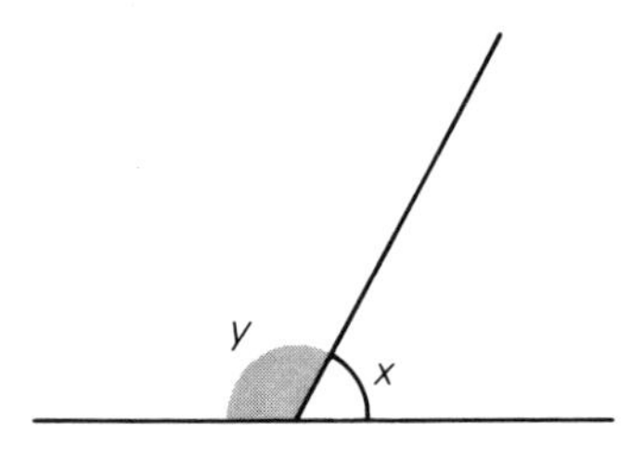

(a) If $\angle x = 50°$, what is $\angle y$?
(b) if $\angle y = 2x$, what is $\angle x$?
(c) if $y = x + 50°$, what is $\angle x$?

14 Statistics

Pictogram

Pictograms are used for showing facts quickly. They do not show the facts accurately, and some pictograms can give quite the wrong idea. The pictogram below shows grain crops at Cowhorn Farm. Each symbol represents 5 hectares under cultivation. (1 hectare = 10 000 m^2, which is about the size of a football field.)

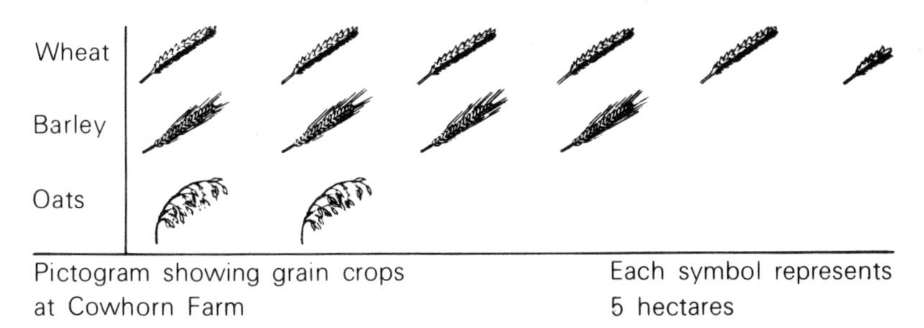

Pictogram showing grain crops at Cowhorn Farm

Each symbol represents 5 hectares

Pie Chart

Pie charts are used to show percentages. The pie chart below shows the area of each part of the United Kingdom as a percentage of the whole country. Notice that it is the percentage which is marked in each case, and not the area, nor the angle of each 'slice'.

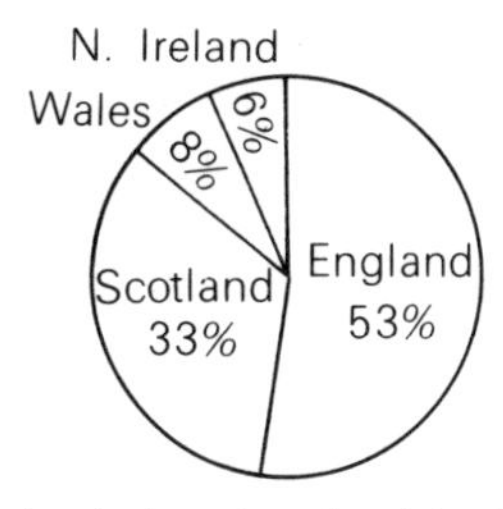

Pie chart showing land in the UK

Bar Chart

Bar charts can be used to show rainfall, imports and exports and so on. The bar chart below shows the increase in vehicles on the roads of the United Kingdom from 1930 to 1970.

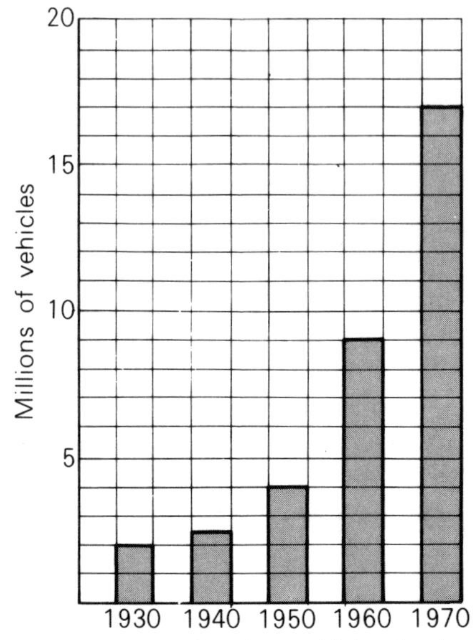

Bar chart showing vehicles in the UK

Continuous Graph

The continuous graph below shows the increase in a boy's height over a period of six years. His height passes through all possible values from 135 cm to 168 cm. A smooth curve can be drawn through the points, and this gives a good idea of the boy's height at in-between times.

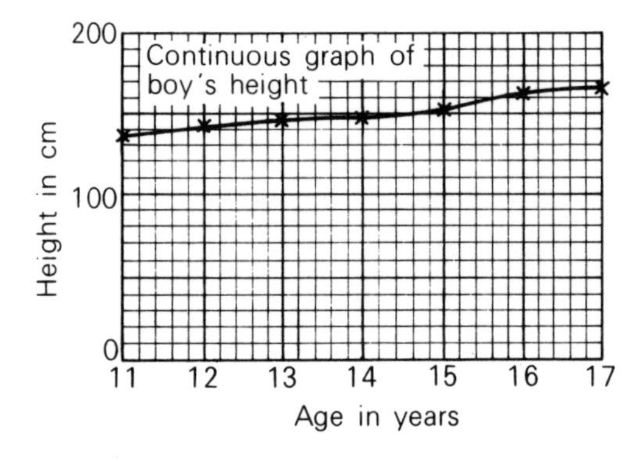

Line Graph

A line graph shows how things vary, but the points are joined by straight lines. This is because the points which are plotted tell us nothing about the in-between values. On the line graph below, a patient in hospital has had his temperature taken at noon every day. The graph does not tell us his temperature at midnight.

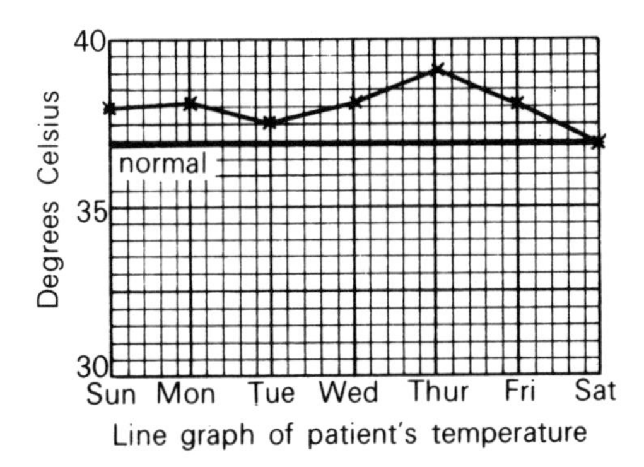

Line graph of patient's temperature

Exercise 14a

Questions 1–15 refer to the diagrams on pages 65–7.

(1) Which is the largest grain crop at Cowhorn Farm?
(2) Which is the smallest crop at Cowhorn Farm?
(3) How many hectares are given over to barley?
(4) How many vehicles were there in the United Kingdom in 1930?
(5) How many vehicles were there in the United Kingdom in 1970?
(6) In which ten year period was there the greatest increase in vehicles?
(7) Which is the smallest part of the United Kingdom?
(8) What percentage of the land area of the United Kingdom is Scotland?
(9) How old was the boy during the period in which his height was measured?
(10) In which year did he grow most in this time?
(11) In which year was there least growth?
(12) On which day was the patient's temperature reading highest?
(13) On which day was the patient's temperature reading lowest?
(14) On which day was the patient's temperature reading normal?
(15) On which day was the patient's temperature highest?
(16) Find how many people in your mathematics class walk to school. Find how many come by car, bus, or bicycle. Illustrate your findings by drawing a pictogram.

(17) Out of 100 applicants for jobs with a civil airline, 43 wanted to be pilots, 25 wanted to be radio operators, 17 wanted to be navigators, and the rest were mechanics. Show these facts using a pictogram.

(18) Draw a bar chart to show the monthly rainfall at Edinburgh, as shown in the table.

Month	Jan	Feb	March	April	May	June
Rainfall in cm	6.3	4.3	4.1	4.1	5.6	4.8

Month	July	Aug	Sept	Oct	Nov	Dec
Rainfall in cm	7.6	8.1	6.6	7.1	6.1	5.3

(19) The table shows the growth of population in Scotland in the nineteenth century. The figures are in millions.

	1821	1831	1841	1851	1861
Men	1.0	1.1	1.2	1.4	1.5
Women	1.1	1.3	1.4	1.5	1.6
Total	2.1	2.4	2.6	2.9	3.1

	1871	1881	1891	1901
Men	1.6	1.8	1.9	2.2
Women	1.8	1.9	2.1	2.3
Total	3.4	3.7	4.0	4.5

Illustrate the growth in population by drawing a pictogram.

(20) Using the figures given in question 19, illustrate the growth of the population using a bar chart.

(21) Draw a bar chart to compare the growth of population of men and women in Scotland. Use the figures given in question 19.

(22) The table shows the lengths, in kilometres, of the five longest rivers in North America.

River	Length (km)
Mississippi—Missouri	6 051
Mackenzie	4 241
Yukon	3 185
St Lawrence	3 130
Rio Grande	3 034

Draw a horizontal bar chart to show these lengths.

(23) The table shows average hours of daily sunshine in London, for each month of the year.

Month	Jan	Feb	March	April	May	June
Hours of sunshine	1.1	1.8	3.4	4.4	5.9	6.6

Month	July	Aug	Sept	Oct	Nov	Dec
Hours of sunshine	6.2	5.8	4.5	3.1	1.4	0.9

Draw a line graph to illustrate these figures.

(24) Draw a line graph to illustrate the rainfall given in question 18.

(25) Draw a line graph to illustrate the increase in population in Scotland, as shown in question 19.

(26) Draw a pie chart to show that 71% of the earth's surface is covered with water.

(27) The atmosphere consists of 76% nitrogen, 20% oxygen, and 4% other gases. Show these facts on a pie chart.

(28) Draw a pie chart to illustrate the facts which you obtained for question 16.

(29) Draw a pie chart to illustrate the facts given in question 17.

(30) Illustrate the facts given in question 23 by using a bar chart.

Exercise 14b

(1) The diagram shows the increase in world population from 1700 to 1900.

1900

1850

1800

1750

1700

(a) What is the name for this kind of diagram?

(b) If the population in 1700 was 600 million, how many people does each symbol represent?

(c) What was the population in 1900?

(d) The population in the year 2000 is expected to be 6 000 million. How many symbols would be needed to represent this population?

(2) The table gives the mean average monthly rainfall at Cape Town.

Month	Jan	Feb	March	April	May	June
Rainfall (mm)	15	8	18	48	79	84

Month	July	Aug	Sept	Oct	Nov	Dec
Rainfall (mm)	89	66	43	30	18	10

(a) Draw a bar chart to illustrate these figures.
(b) What is the total annual rainfall at Cape Town?
(c) In which 3-monthly period is there most rain?
(d) In which 3-monthly period is there least rain?

(3) In a class of 36 pupils there are 20 boys.
(a) Draw a pie chart to illustrate this.
(b) What angle have you used to represent each pupil?
(c) What angle have you used to represent the girls?

(4) The table shows the growth of a girl from 9–15 years.

Age (yr)	9	10	11	12	13	14	15
Height (cm)	135	140	153	157	160	163	165

(a) Draw a graph of this data.
(b) At what age was growth most rapid?
(c) From your graph, find an approximate value for the girl's height at $10\frac{1}{2}$ years.

(5) The table shows average monthly temperatures at Oxford.

Month	Jan	Feb	March	April	May	June
Temp (°C)	4	5	7	9	12	15

Month	July	Aug	Sept	Oct	Nov	Dec
Temp (°C)	17	17	14	11	5	4

(a) Draw a line graph to show the temperature variation.
(b) Calculate the average of the monthly readings.
(c) In which month do the hottest days occur?

Section C
15 Number Patterns

Rectangle Numbers

A number can be shown as a set of dots. If the dots can be put into two or more equal rows, the number is a *rectangle number*. Any number which has factors (other than itself and 1) is a rectangle number. A rectangle number is a *multiple*.

It may be possible to make different rectangle patterns of the same number. Different rectangle patterns show different sets of factors.

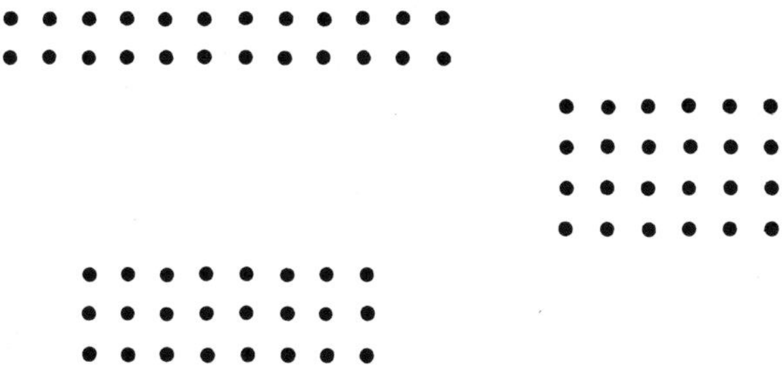

A number which is not a rectangle number is a *prime number*.

Square Numbers

Some rectangle numbers can be put as a square pattern. These are the *square numbers*. A square number is sometimes called a *perfect square*.

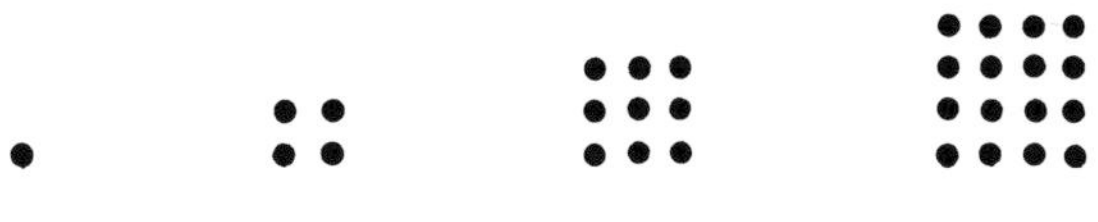

Triangle Numbers

Some numbers can be put as triangle patterns.

Two sets of triangle numbers can be put together to make a set of rectangle numbers. Or they can be put together to make a set of square numbers.

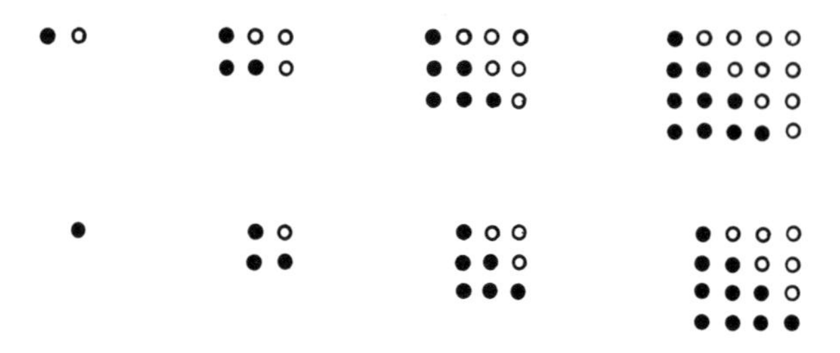

Patterns of Numerals

By splitting up square patterns we can obtain interesting patterns of numerals.

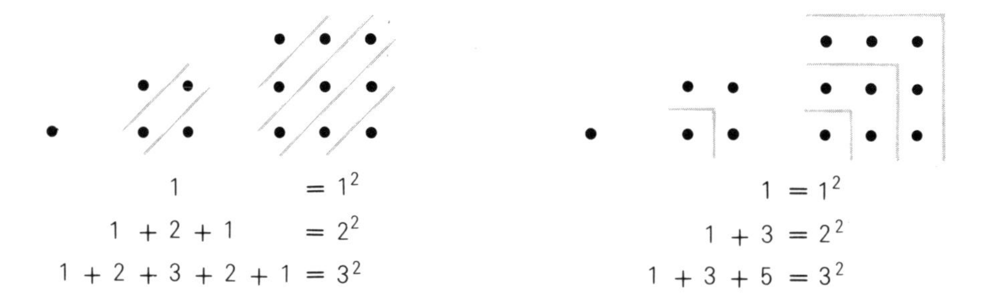

$$1 = 1^2$$
$$1 + 2 + 1 = 2^2$$
$$1 + 2 + 3 + 2 + 1 = 3^2$$

$$1 = 1^2$$
$$1 + 3 = 2^2$$
$$1 + 3 + 5 = 3^2$$

Sequences

The numbers

1, 2, 3, 4, 5, . . .

are the first five natural numbers. It is quite easy to see that the next number should be 6. This is because we can see a connection between a number and the one before it.

2, 5, 8, 11, 14, ...

is much the same. We can see that each number is three more than the one before it, and we can see that the next number must be 17. A set of numbers in which we can work out what the next number is going to be is called a *sequence*.

You should be familiar with the sequence of triangle numbers, and square numbers.

1, 3, 6, 10, ...
1, 4, 9, 16, ...

Can you find the next number in each of the following?

1, 2, 4, 8, ...
1, 3, 9, 27, ...
1, 8, 27, 64, ...

Fibonacci Sequence

This sequence is one in which each term is found by adding the two terms just before it.

1, 1, 2, 3, 5, 8, 13, ...

Exercise 15a

Write the next three terms in each of the sequences in questions 1–20.

(1) 1, 2, 3, 4, 5, 6, 7, 8, 9, 10,
(2) 2, 6, 10, 14, 18,
(3) 10, 8, 6, 4, 2, 0, -2,
(4) 15, 10, 5,
(5) 1, 2, 4, 8, 16,
(6) 3, 6, 12, 24,
(7) 2, 6, 18, 54,
(8) 8, 4, 2, 1, $\frac{1}{2}$,
(9) 1, 1, 2, 3, 5, 8,
(10) 1, 3, 6, 10,
(11) 1, 4, 9, 16,
(12) 1, 8, 27, 64,
(13) 2, 6, 12, 20,
(14) $\frac{1}{2}, \frac{2}{3}, \frac{3}{4}, \frac{4}{5}$,
(15) $\frac{1}{3}, \frac{3}{5}, \frac{5}{7}, \frac{7}{9}$,
(16) 3, 7, 13, 21,
(17) 1, 3, 9, 27,
(18) $\frac{1}{2}, \frac{1}{4}, \frac{1}{8}, \frac{1}{16}$,
(19) 625, 125, 25,
(20) $\frac{1}{8}, \frac{1}{4}, \frac{1}{2}$,

(21) Draw four different rectangle patterns to show the number 48.
(22) Draw five different rectangle patterns to show the number 60.
(23) Draw four different rectangle patterns to show the number 36.
(24) Draw square patterns to illustrate the first seven square numbers.
(25) By drawing a dot pattern, find the 10th triangle number.

(26) Write down all the prime numbers between 10 and 30.
(27) Write down powers of 2 from 2^1 to 2^8.
(28) Write down powers of 3 from 3^1 to 3^6.
(29) Write down powers of 4 from 4^1 to 4^4.
(30) Write down powers of 5 from 5^1 to 5^4.

Exercise 15b

(1) Make up a table of powers of 2 up to 2^{10}, and use the table to calculate:
(a) $2^8 - 2^5$,
(b) $\dfrac{1\,024 \times 128}{32 \times 512}$
(c) $\dfrac{\sqrt{256 \times 512}}{128}$
(d) 4^4.
(2) (a) Draw dot patterns to show the triangle numbers 1, 3, 6, 10.
(b) Write down the first ten triangle numbers.
(c) Show how to combine two sets of triangle numbers to make up the square numbers.
(d) Show how to combine two sets of triangle numbers to make the rectangle numbers 2, 6, 12, 20.
(3) The number pattern shown is known as Pascal's triangle.

$$\begin{array}{ccccccccc} & & & & 1 & & & & \\ & & & 1 & & 1 & & & \\ & & 1 & & 2 & & 1 & & \\ & 1 & & 3 & & 3 & & 1 & \\ 1 & & 4 & & 6 & & 4 & & 1 \end{array}$$

(a) Write down the next row of figures.
(b) Which row starts 1, 10, 45?
(c) Add up the numbers in each of the rows shown. What is the sum of the numbers in the tenth row?
(4) $1^2 - 0^2 = 1$
$2^2 - 1^2 = 3$
$3^2 - 2^2 = 5$
$4^2 - 3^2 = 7$

(a) Copy this number pattern and add the next three rows.
(b) Write down the twentieth line in the pattern.
(c) Write down the value of $123^2 - 122^2$.

16 Drawing Triangles

Given Three Sides (AB, BC, CA)

(1) Draw AB.
(2) With centre A, draw an arc equal to AC.
(3) With centre B, draw an arc equal to BC.
(4) The arcs cut at C. Join AC and BC.

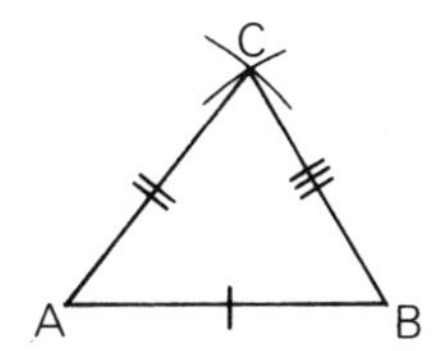

Two Sides and the Included Angle (AB, AC, ∠A)

(1) Draw AB.
(2) Using your protractor, draw ∠A.
(3) With centre A, draw an arc with radius equal to AC.
(4) Join BC.

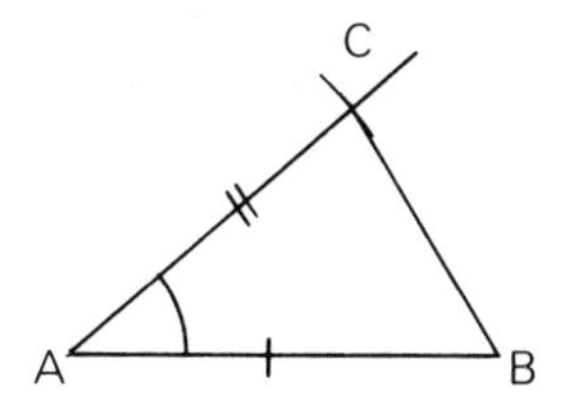

Two Angles and One Side (AB, ∠A, ∠B)

(1) Draw AB.
(2) Draw ∠A.

(3) Draw ∠B. (This would have to be calculated if it is not given.)

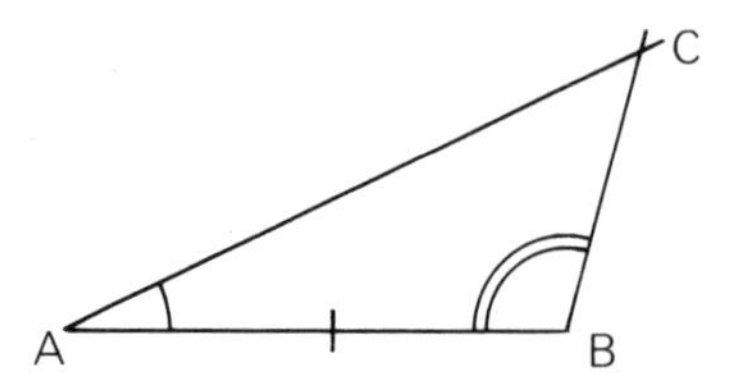

Right Angle, Hypotenuse and Side (AB, BC, ∠A = 90°)

(1) Draw AB.
(2) Draw the right angle at A.
(3) With centre B, draw an arc equal to BC.
(4) Draw BC.

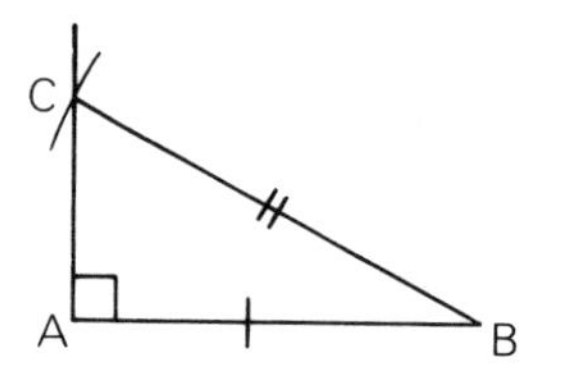

Drawing Quadrilaterals

A triangle is a rigid shape. A quadrilateral is not. If you are given the four sides of a quadrilateral, there are an infinite number of quadrilaterals which could be drawn. All three quadrilaterals shown, have sides 1, 2, 3 and 4 cm.

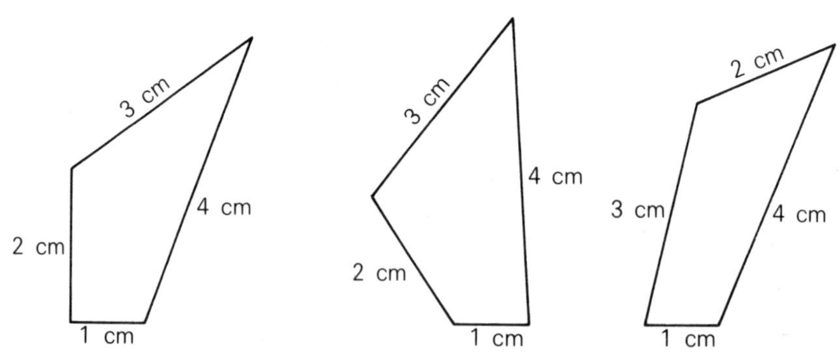

Before drawing a quadrilateral, it is necessary to know an angle or a diagonal. If you are given four sides and one diagonal, this is the same as two triangles, each of which have three sides given. In general, five facts are needed before a quadrilateral can be drawn.

Example: Draw the quadrilateral ABCD, where AB = 2 cm, BC = 3 cm, CD = 3 cm, DA = 2 cm, BD = 3 cm.

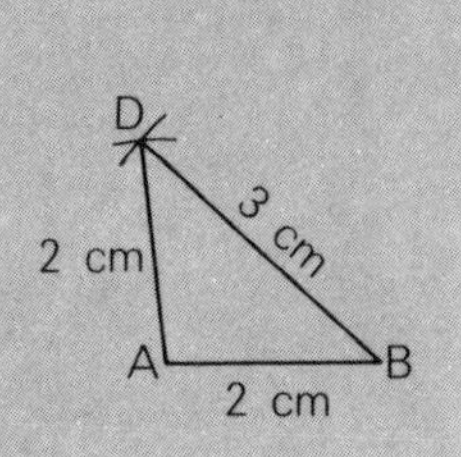

1. Draw △ABD.

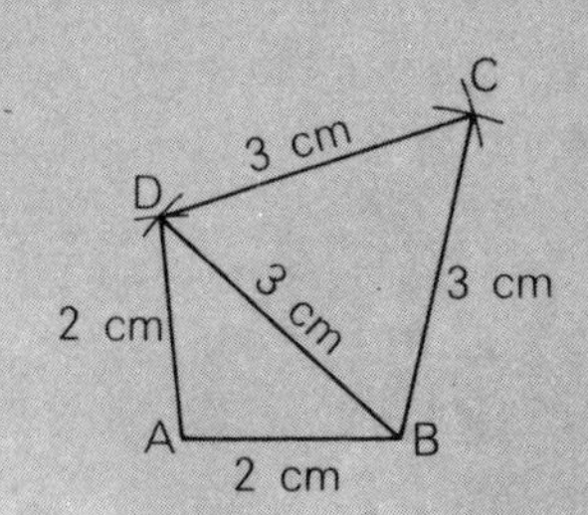

2. Draw △BCD.

Example: Draw the quadrilateral ABCD, where BC = 4 cm, AB = AD = 3 cm, AC = 5.5 cm and ∠DAB = 70°.

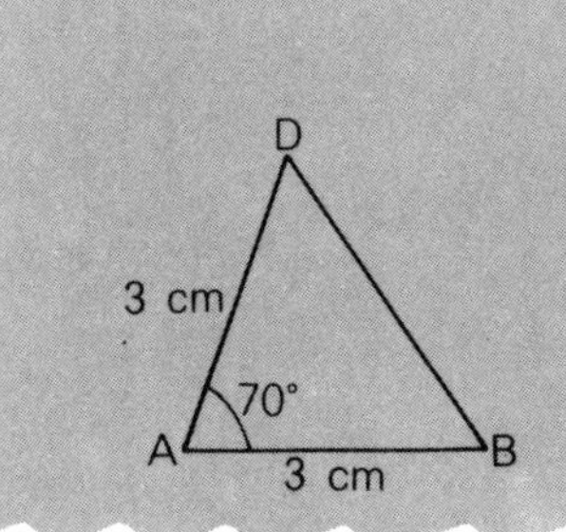

1. Draw △ABD.

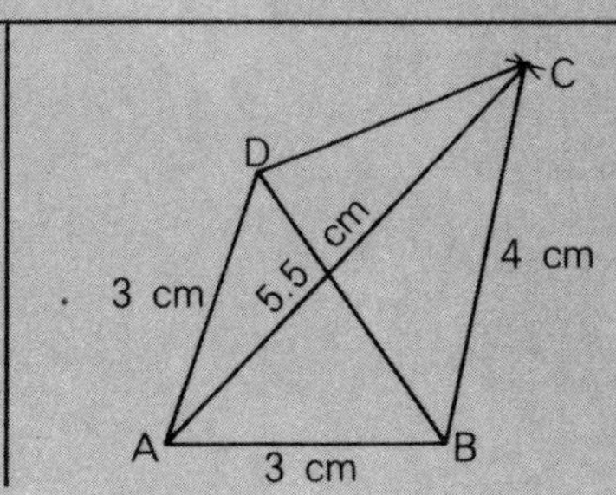

2. Draw △ABC and join CD.

Remember when lettering your quadrilateral that the letters A, B, C and D go round the figure. The diagonals are AC and BD. Always start by making a rough sketch.

Exercise 16a

Draw triangles from the following data:

(1) AB = 8 cm, BC = 6 cm, AC = 7 cm. Measure ∠ABC.
(2) PQ = 6 cm, QR = 3 cm, PR = 7.5 cm. Measure ∠PQR.
(3) XY = 8 cm, YZ = 4 cm, ∠XYZ = 50°. Measure XZ.
(4) KL = 9.5 cm, ∠MKL = 50°, ∠MLK = 60°. Measure KM.
(5) PQ = 12.4 cm, PR = 12 cm, ∠PRQ = 90°. Measure QR.
(6) AC = 10 cm, ∠CBA = 80°, ∠ACB = 35°. Measure AB.
(7) AB = 8 cm, BC = 5 cm, ∠ABC = 130°. Measure AC.
(8) LM = 6 cm, KL = KM = 8 cm. Measure ∠KLM.

(9) XY = 7 cm, XZ = 8 cm, $\angle$XYZ = 90°. Measure $\angle$YXZ.
(10) AC = 6 cm, $\angle$BAC = 120°, $\angle$ACB = 20°. Measure BC.
(11) PQ = 7.5 cm, RQ = 6 cm, $\angle$PQR = 60°. Measure PR.
(12) RS = 14.6 cm, ST = 15 cm, $\angle$RST = 90°. Measure RT.
(13) AB = 5.5 cm, BC = 5 cm, AC = 9 cm. Measure $\angle$ABC.
(14) XZ = 6 cm, YZ = 14 cm, $\angle$ZXY = 90°. Measure $\angle$XZY.
(15) XZ = 6 cm, YZ = 8 cm, $\angle$XZY = 90°. Measure XY.
(16) PQ = 5.5 cm, $\angle$RPQ = $\angle$RQP = 40°. Measure PR.
(17) AC = 7 cm, BC = 9.9 cm, $\angle$BAC = 90°. Measure $\angle$ABC
(18) XY = 12 cm, XZ = 12.2 cm, YZ = 2.2 cm. Measure $\angle$XYZ.
(19) PQ = 4.5 cm, $\angle$PQR = 130°, $\angle$PRQ = 25°. Measure QR.
(20) KL = 6 cm, KM = 6 cm, $\angle$LKM = 60°. Measure LM.

Draw quadrilaterals from the following data:

(21) AB = 7 cm, BC = 8 cm, AC = 9 cm, AD = CD = 6 cm. Measure BD.
(22) AB = 8 cm, BC = 7 cm, CD = 5 cm, $\angle$ABC = 40°, $\angle$BCD = 25°. Measure AC.
(23) PR = 8 cm, PS = 10 cm, $\angle$PRS = 90°, $\angle$QPR = 40°, $\angle$QRP = 70°. Measure QS.
(24) WX = 9 cm, WY = 10 cm, WZ = 11 cm, $\angle$WXY = $\angle$WYZ = 90°. Measure XZ.
(25) PQ = QS = PS = 10 cm, PR = 12 cm, RS = 8 cm. Measure $\angle$QRS.

Exercise 16b

(1) If AB = 9 cm, AD = 8 cm, BD = 6 cm, (a) draw $\triangle$ABD, (b) measure $\angle$BAD, (c) if AC = 10 cm and CD = 4 cm, complete the quadrilateral ABCD, (d) measure BC.
(2) If QS = 7 cm, $\angle$RQS = 30°, $\angle$RSQ = 50°, (a) draw $\triangle$QRS, (b) measure RS, (c) complete the quadrilateral PQRS by making $\angle$PQR = 80°, PQ = 8.3 cm, (d) measure PS.
(3) $\triangle$UST has US = 7.5 cm, ST = 8 cm, $\angle$UST = 35°, (a) draw $\triangle$UST, (b) measure $\angle$SUT, (c) complete the quadrilateral RSTU, where RT = 7 cm, $\angle$STR = 50°, (d) measure RU.
(4) If WZ = 8 cm, WY = 10 cm, and $\angle$YZW = 90°, (a) draw $\triangle$WYZ, (b) measure YZ, (c) complete the quadrilateral WXYZ, where XY = 6 cm, XW = 8 cm, (d) measure $\angle$YXW.
(5) $\triangle$LMN is such that LN = 8 cm, $\angle$MLN = $\angle$MNL = 62°, (a) draw $\triangle$LMN, (b) measure MN, (c) complete the quadrilateral KLMN, where $\angle$KMN = 35° and $\angle$LNK = 20°, (d) measure CK.

17 Metric Weights and Measures

The Metric System

The metric system was made up in France during the French Revolution. It took the place of the old French system of units. Since that time, the metric system has come into use in countries other than France, but there were slight differences from place to place. In 1960 an international conference agreed on the metric units to be used, and on the short forms for each unit. These units which are now used by many countries are called SI units. The letters SI are the first letters of the French words for 'international system'.

The units given in the following tables are metric units which are often used. Many of them are SI units. Some of them are not.

Length

Apart from the metre (m), the most common units are the millimetre (mm), the centimetre (cm) and the kilometre (km).

10 mm = 1 cm
100 cm = 1 m
1 000 mm = 1 m
1 000 m = 1 km

Area

100 mm^2 = 1 cm^2
10 000 cm^2 = 1 m^2
10 000 m^2 = 1 ha (hectare)
100 ha = 1 km^2

It is useful to remember that one hectare is slightly larger than a football pitch.

Volume

The cubic metre (m^3) and cubic centimetre (cm^3) are both used to measure volume. For liquids, the litre (l) and millilitre (ml) are used.

$$1 \text{ ml} = 1 \text{ cm}^3$$
$$1\,000 \text{ ml} = 1 \text{ l}$$
$$1\,000 \text{ l} = 1 \text{ m}^3.$$

Mass

The units of mass are the milligram (mg), the gram (g), the kilogram (kg) and the tonne (t).

$$1\,000 \text{ mg} = 1 \text{ g}$$
$$1\,000 \text{ g} = 1 \text{ kg}$$
$$1\,000 \text{ kg} = 1 \text{ t}.$$

Water

1 ml (or 1 cm^3) of water weighs 1 g
1 l of water weighs 1 kg
1 m^3 of water weighs 1 t.

Speed

Speed is usually given in kilometres per hour (km/h), or metres per second (m/s) or centimetres per second (cm/s).

$$100 \text{ cm/s} = 1 \text{ m/s}$$
$$= 3.6 \text{ km/h}.$$

Fuel

Fuel consumption is given in kilometres per litre (km/l).

Exercise 17a

Write in centimetres:

(1) 3 m 2 cm (2) 14 m 5 cm (3) 165 mm

Write in metres:

(4) 173 cm (5) 3 km 154 m (6) 736 cm

Write in square metres:

(7) 1 362 cm^2 (8) 2 536 cm^2 (9) 3.2 ha

Write in hectares:

(10) 17 000 m^2 (11) 35 km^2 (12) 4.3 km^2

Write in litres:

(13) 3 000 ml (14) 5 200 ml (15) 700 ml

Write in millilitres:

(16) 4 litres (17) 2.5 litres (18) 0.3 litres

Write in grams:

(19) 5 000 mg (20) 3.2 kg (21) 4.7 kg

Write in kilograms:

(22) 7 500 g (23) 4 620 g (24) 5.3 t

In questions 25–35, add the amounts given.

(25) 5.25 m, 3.02 m, 53 cm, 246 cm and 20 cm. Give your answer in metres.
(26) 7.3 cm, 44 mm, 27 mm, 8.4 cm and 17.35 cm. Give your answer in centimetres.
(27) 8.91 cm, 2.51 cm, 83 mm, 2 mm, and 4.1 cm. Give your answer in millimetres.
(28) 453 m, 3 470 m, 2.3 km, 7.2 km, and 436 m. Give your answer in kilometres.
(29) 0.25 m^2, 5 200 cm^2, 1.3 m^2, 9 700 cm^2 and 7 400 cm^2. Give your answers in square metres.
(30) 5 900 m^2, 3 200 m^2, 8 400 m^2, 7 200 m^2 and 3.56 ha. Give your answer in hectares.
(31) 576 ha, 27 ha, 934 ha, 2.7 km^2 and 3.8 km^2. Give your answer in square kilometres.
(32) 5 976 ml, 3 428 ml, 6 027 ml, 4.7 l and 8.3 l. Give your answer in litres.
(33) 597 g, 382 g, 4 962 g, 877 g, 5.73 kg and 2.73 kg. Give your answer in kilograms.
(34) 624 mg, 727 mg, 5.3 g, 7.8 g and 42.3 g. Give your answer in grams.

(35) 598 kg, 674 kg, 5 736 kg, 8.7 t and 9.3 t. Give your answer in tonnes.

Exercise 17b

(1) A rectangular tank is 2.3 m long, 97 cm wide and 1.2 m deep.

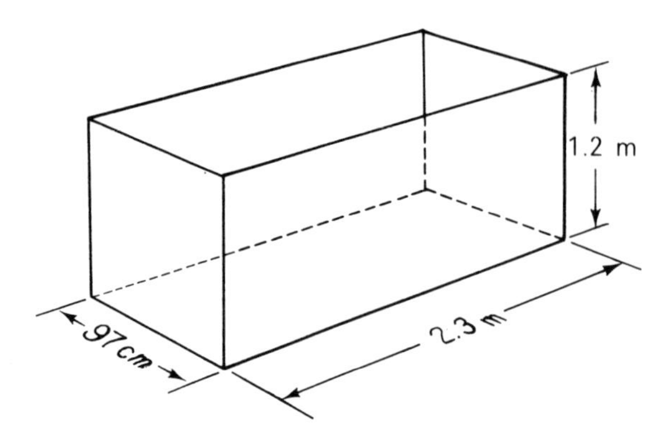

Find:

(a) the volume of the tank in m^3 to one decimal place ($V = lbh$),

(b) the weight of acid needed to fill the tank if 1 m^3 of acid weighs 1.2 t (give your answer in t to one decimal place),

(c) how many carboys of acid would be needed in order to fill the tank, if each carboy holds 50 litres?

(2) A football pitch is 100 m long and 75 m wide. Find:

(a) the length of the perimeter of the pitch in metres,

(b) the exact area of the pitch in hectares,

(c) the cost of turfing the pitch at 22p per m^2.

(3) A car travels at an average speed of 45 km/h with a fuel consumption of 12 km/l. Find:

(a) the distance travelled in 3 hours 20 minutes,

(b) the fuel used in the same time,

(c) how far the car would travel on a full tank of 33 l,

(d) the cost of a full tank of petrol at 16p a litre.

(4) A metal sphere has a radius of 21 cm. Find:

(a) the surface area of the sphere in cm^2 ($A = \frac{88r^2}{7}$),

(b) the volume of the sphere in cm^3 ($V = \frac{88r^3}{21}$),

(c) the weight of the sphere, correct to the nearest kg, if the metal weighs 8.9 g/cm^3.

18 Fractions

Types of Fractions

There are three names given to fractions. These are shown below.

$\frac{3}{7}$	$\frac{9}{8}$	$1\frac{2}{5}$
Vulgar fraction	Improper fraction	Mixed number

In any fraction, the top number is called the *numerator*. The bottom number is called the *denominator*.

Equivalent Fractions

If two fractions have the same value, they are *equivalent fractions*.

$$\frac{3}{4} = \frac{6}{8} = \frac{9}{12}.$$

These are equivalent fractions.

If the top and bottom of a fraction are both multiplied by the same number, the value of the fraction is not changed.

$$\frac{5}{7} = \frac{5 \times 2}{7 \times 2} = \frac{10}{14}.$$

In the same way, the top and bottom of a fraction can be divided by the same number without changing its value.

$$\frac{\overset{2}{\cancel{8}}}{\underset{3}{\cancel{12}}} = \frac{2}{3}.$$

This is called *cancelling*.

Ordering

To find which of two fractions is the larger, they both must have the same denominator.

Example: Which is larger, $\frac{7}{8}$ or $\frac{11}{12}$?

24 is a multiple of 8 and 12	$\frac{7}{8} = \frac{21}{24}$ $\frac{11}{12} = \frac{22}{24}$	$\frac{11}{12} > \frac{7}{8}$
1. Find a number which is a multiple of 8 and 12.	2. Change $\frac{7}{8}$ and $\frac{11}{12}$ to 24ths.	3. Because $\frac{22}{24} > \frac{21}{24}$, $\frac{11}{12}$ is the larger fraction.

Addition and Subtraction

In order to add or subtract fractions, they must have the same denominator.

Examples:

1. Change the fractions so that they have the same denominator.
2. Add.

$$\frac{3}{7} + \frac{2}{5} = \frac{15}{35} + \frac{14}{35} = \frac{29}{35}.$$

1. Change the fractions so that they have the same denominator.
2. Subtract.
3. Cancel.

$$\frac{11}{12} - \frac{2}{3} = \frac{11}{12} - \frac{8}{12} = \frac{3}{12} = \frac{1}{4}.$$

1. Change the fractions so that they have the same denominator.
2. Add and subtract.

$$\frac{1}{2} + \frac{2}{5} - \frac{4}{7} = \frac{35}{70} + \frac{28}{70} - \frac{40}{70} = \frac{23}{70}.$$

When there are mixed numbers, add and subtract the whole numbers first.

Example: $1\frac{3}{12} - 2\frac{1}{2} + 3\frac{1}{6}$

1. Add and subtract the whole numbers.

$$1\frac{3}{12} - 2\frac{1}{2} + 3\frac{1}{6} = 2 + \frac{3}{12} - \frac{1}{2} + \frac{1}{6}$$

2. Change the fractions so that they have the same denominator.

$$= 2 + \frac{3}{12} - \frac{6}{12} + \frac{2}{12}$$

3. Add and subtract the fractions.

$$= 2 - \frac{1}{12}$$

4. Subtract the fraction from the whole number.

$$= 1\tfrac{11}{12}.$$

Multiplication

To multiply fractions, multiply the numerators, then multiply the denominators. Mixed numbers must first be changed to improper fractions.

Example: $1\frac{3}{4} \times 2\frac{1}{2}$

1. Change mixed numbers to improper fractions.

$$1\frac{3}{4} \times 2\frac{1}{2} = \frac{7}{4} \times \frac{5}{2}$$

2. Multiply.

$$= \frac{7 \times 5}{4 \times 2}$$

$$= \frac{35}{8}$$

3. Change back to mixed numbers.

$$= 4\tfrac{3}{8}.$$

Cancelling

When multiplying fractions, always cancel when you can. This will save you a lot of hard work.

Example: $6\frac{3}{7} \times 5\frac{4}{9}$

First method:	Second method:
$6\frac{3}{7} \times 5\frac{4}{9} = \frac{45}{7} \times \frac{49}{9}$	$6\frac{3}{7} \times 5\frac{4}{9} = \frac{\cancel{45}^{5}}{\cancel{7}_{1}} \times \frac{\cancel{49}^{7}}{\cancel{9}_{1}}$
$= \frac{2\,205}{63}$	$= 35.$
$= 35.$	

The example shows how much easier the second method is than the first. In the first method there is a long multiplication and a difficult division. In the second method the whole thing is much easier.

Division

If the product of two fractions is 1, then each fraction is the *inverse* of the other. For example, $\frac{2}{3} \times \frac{3}{2} = 1$, so that $\frac{2}{3}$ is the inverse of $\frac{3}{2}$. Also $\frac{3}{2}$ is the inverse of $\frac{2}{3}$.

Dividing by a fraction is the same as multiplying by its inverse. Mixed numbers must first be changed to improper fractions.

Examples:

1. Find the inverse of the divisor, and change to multiplication.

$$\frac{3}{5} \div \frac{2}{7} = \frac{3}{5} \times \frac{7}{2}$$

2. Multiply.

$$= \frac{21}{10}$$

3. Change to mixed numbers.

$$= 2\frac{1}{10}.$$

1. Change mixed numbers to improper fractions.

$$1\frac{1}{2} \div 2\frac{1}{4} = \frac{3}{2} \div \frac{9}{4}$$

2. Find the inverse of the divisor and change to multiplication. Cancel.

$$= \frac{\cancel{3}^{1}}{\cancel{2}_{1}} \times \frac{\cancel{4}^{2}}{\cancel{9}_{3}}$$

3. Multiply.

$$= \frac{2}{3}.$$

Exercise 18a

In questions 1–10, write the fractions in order of increasing size.

(1) $\frac{4}{7}, \frac{2}{3}, \frac{3}{5}$
(2) $\frac{5}{11}, \frac{2}{5}, \frac{3}{7}$
(3) $\frac{2}{3}, \frac{5}{8}, \frac{3}{4}$
(4) $\frac{2}{9}, \frac{1}{6}, \frac{1}{5}$
(5) $\frac{9}{10}, \frac{4}{5}, \frac{7}{8}, \frac{3}{4}$
(6) $\frac{4}{11}, \frac{3}{7}, \frac{7}{8}, \frac{2}{3}$
(7) $\frac{5}{8}, \frac{3}{4}, \frac{4}{7}, \frac{2}{3}$
(8) $\frac{2}{5}, \frac{5}{9}, \frac{4}{7}, \frac{1}{3}$
(9) $\frac{3}{4}, \frac{4}{5}, \frac{1}{2}, \frac{3}{7}$
(10) $\frac{7}{10}, \frac{5}{8}, \frac{6}{7}, \frac{2}{3}$
(11) $2\frac{1}{2} + 3\frac{2}{3}$
(12) $3\frac{1}{4} + 2\frac{2}{7}$
(13) $1\frac{1}{5} + 4\frac{2}{3}$
(14) $5\frac{3}{7} + 3\frac{4}{5}$
(15) $4\frac{3}{4} + 5\frac{5}{6}$
(16) $4\frac{3}{5} - 2\frac{1}{2}$
(17) $3\frac{3}{4} - 2\frac{2}{7}$
(18) $5\frac{4}{9} - 3\frac{2}{3}$
(19) $6\frac{1}{3} - 3\frac{2}{5}$
(20) $4\frac{1}{2} - 3\frac{7}{9}$
(21) $1\frac{2}{3} + 3\frac{3}{4} - 2\frac{1}{2}$
(22) $2\frac{4}{5} - 1\frac{1}{2} + 3\frac{1}{10}$
(23) $4\frac{1}{3} + 2\frac{2}{5} - 3\frac{1}{2}$
(24) $6\frac{1}{4} + 2\frac{1}{2} - 3\frac{1}{3}$
(25) $2\frac{4}{7} - \frac{3}{4} + 1\frac{1}{2}$
(26) $3\frac{1}{2} + 4\frac{7}{12} - \frac{1}{5}$
(27) $2\frac{1}{3} + 3\frac{1}{4} - 2\frac{2}{3}$
(28) $2\frac{1}{2} + 3\frac{5}{24} - 2\frac{3}{4}$
(29) $4\frac{1}{2} - 3\frac{2}{3} + 5\frac{1}{8}$
(30) $2\frac{1}{4} + 7\frac{1}{3} - 4\frac{3}{8}$
(31) $\frac{2}{3} \times \frac{5}{7}$
(32) $\frac{1}{2} \times \frac{3}{5}$
(33) $\frac{2}{5} \times \frac{4}{9}$
(34) $\frac{3}{5} \times \frac{2}{3}$
(35) $\frac{2}{9} \times \frac{3}{4}$
(36) $1\frac{2}{3} \times 2\frac{2}{5}$
(37) $2\frac{1}{2} \times 1\frac{1}{4}$
(38) $3\frac{1}{7} \times 1\frac{3}{4}$
(39) $5\frac{2}{5} \times 1\frac{1}{9}$
(40) $2\frac{1}{4} \times \frac{2}{3}$
(41) $\frac{2}{3} \div \frac{1}{2}$
(42) $\frac{3}{4} \div \frac{2}{7}$
(43) $\frac{3}{5} \div \frac{2}{3}$
(44) $\frac{4}{11} \div 1\frac{1}{3}$
(45) $1\frac{3}{4} \div 2\frac{1}{3}$
(46) $1\frac{5}{9} \div \frac{7}{11}$
(47) $2\frac{4}{5} \div \frac{2}{5}$
(48) $2\frac{1}{4} \div 2\frac{1}{2}$
(49) $3\frac{4}{7} \div 1\frac{2}{3}$
(50) $4\frac{1}{5} \div 1\frac{1}{10}$

Exercise 18b

(1) Find the value of,
  (a) $1\frac{1}{2} + 3\frac{2}{5} - 2\frac{1}{4}$,
  (b) $2\frac{1}{2} \times 1\frac{1}{4}$,
  (c) $3\frac{1}{2} \div 2\frac{1}{4}$.
(2) (a) Write in order of increasing size $\frac{3}{7}, \frac{5}{8}, \frac{7}{20}$.
  (b) Find the value of $1\frac{1}{2} + 3\frac{3}{4} - 2\frac{5}{6}$.
  (c) Multiply $\frac{3}{7} \times \frac{5}{6} \times \frac{7}{20}$.
(3) Calculate,
  (a) $5\frac{3}{5} - 1\frac{1}{10}$,
  (b) $1\frac{3}{7} \times 1\frac{1}{10}$,
  (c) $1\frac{3}{7} \div 1\frac{3}{14}$.
(4) Evaluate,
  (a) $1\frac{1}{2} + 2\frac{2}{3}$,
  (b) $3\frac{5}{6} - 2\frac{1}{2}$,
  (c) $\dfrac{1\frac{1}{2} + 2\frac{2}{3}}{3\frac{5}{6} - 2\frac{1}{2}}$.

19 Fractions in Algebra

Numbers and Letters

In algebra, we use letters instead of numbers. The letters must be used in the same way as numbers. Fractions with letters are the same as fractions with numbers. Because we do not know the number that each letter stands for, we cannot find the value of a fraction sum with letters. We can *simplify* an expression which has fractions with letters in it.

Simplification

In order to simplify an expression with fractions, the fractions must all have the same denominator. The fractions can then be added or subtracted in the same way as in chapter 18. When there are whole numbers, remember that

$\frac{a}{a}, \frac{b}{b}, \frac{x}{x}, \frac{x^2}{x^2}, \frac{ab}{ab}$ and so on, are all equal to 1.

Examples:

1. xy is a multiple of x and of y. Change both fractions so that they have xy as denominator.

$$\frac{2}{x} + \frac{3}{y} = \frac{2y}{xy} + \frac{3x}{xy}$$

2. Add.

$$= \frac{2y + 3x}{xy}$$

1. Arrange all fractions and whole numbers so that they have the denominator x^2.

$$1 + \frac{2}{x} - \frac{1}{x^2} = \frac{x^2}{x^2} + \frac{2x}{x^2} - \frac{1}{x^2}$$

2. Add and subtract.

$$= \frac{x^2 + 2x - 1}{x^2}$$

Substitution

You may be able to find the value of a fraction with letters, if you are told the value of each letter.

Example: Find the value of abc if $a = \frac{3}{4}$, $b = 1\frac{2}{5}$, $c = \frac{2}{7}$

1. Substitute the values for a, b, and c.

$$abc = \frac{3}{4} \times 1\frac{2}{5} \times \frac{2}{7}$$

2. Change mixed numbers to improper fractions and cancel.

$$= \frac{3}{\cancel{4}_2} \times \frac{\cancel{7}^1}{5} \times \frac{\cancel{2}^1}{\cancel{7}_1}$$

3. Multiply.

$$= \frac{3}{10}.$$

Exercise 19a

Simplify:

(1) $\frac{a}{2} + \frac{b}{3}$

(2) $\frac{x}{4} + \frac{y}{5}$

(3) $\frac{2}{a} + \frac{3}{b}$

(4) $\frac{3}{x} - \frac{2}{y}$

(5) $\frac{1}{a} - \frac{2}{ab}$

(6) $\frac{3}{b^2} + \frac{4}{b}$

(7) $\frac{3}{4a} - \frac{2}{a}$

(8) $1 + \frac{a}{3}$

(9) $1 - \frac{2}{b}$

(10) $3 - \frac{4}{x^2}$

(11) $2 + \frac{1}{a} + \frac{3}{b}$

(12) $5 - \frac{2}{y}$

(13) $1 + \frac{3x}{4}$

(14) $\frac{3}{x} + 1\frac{1}{2}$

(15) $\frac{2}{ab} + \frac{1}{bc} - \frac{3}{ac}$

Find the value of:

(16) abc if $a = \frac{1}{2}$, $b = \frac{5}{7}$, $c = \frac{2}{5}$
(17) xyz if $x = \frac{3}{11}$, $y = \frac{5}{6}$, $z = 2\frac{3}{4}$
(18) pqr if $p = 1\frac{2}{3}$, $q = 2\frac{2}{5}$, $r = \frac{3}{11}$
(19) lmn if $l = 3\frac{1}{2}$, $m = 1\frac{1}{4}$, $n = 3\frac{1}{7}$
(20) $a + b + c$ if $a = 3\frac{1}{4}$, $b = 2\frac{1}{2}$, $c = 1\frac{1}{8}$
(21) $a + b - c$ if $a = 4\frac{1}{6}$, $b = 2\frac{1}{2}$, $c = 3\frac{1}{3}$
(22) $x - y + z$ if $x = 3\frac{1}{7}$, $y = 2\frac{1}{2}$, $z = 4\frac{1}{4}$
(23) $p + q - r$ if $p = 3\frac{1}{4}$, $q = 2\frac{1}{3}$, $r = 1\frac{5}{6}$
(24) $a - b + c$ if $a = 4\frac{1}{3}$, $b = 3\frac{5}{8}$, $c = 4\frac{1}{4}$
(25) $a \div b$ if $a = 2\frac{7}{10}$, $b = 1\frac{4}{5}$

(26) $\dfrac{x}{y}$ if $x = 3\frac{7}{9}$, $y = 2\frac{5}{6}$

(27) $\dfrac{pq}{r}$ if $p = 3\frac{1}{2}$, $q = 1\frac{1}{3}$, $r = 3\frac{1}{9}$

(28) $\dfrac{ab}{c}$ if $a = 2\frac{1}{7}$, $b = 3\frac{3}{4}$, $c = 12\frac{1}{2}$

(29) $\dfrac{x}{yz}$ if $x = 7\frac{1}{2}$, $y = 3\frac{1}{2}$, $z = 2\frac{1}{4}$

(30) $\dfrac{p}{qr}$ if $p = 7\frac{13}{16}$, $q = 2\frac{1}{2}$, $r = 2\frac{1}{2}$

Exercise 19b

(1) Simplify,
  (a) $\dfrac{x}{3} - \dfrac{y}{4}$,
  (b) $\dfrac{3}{x} - \dfrac{y}{4}$,
  (c) $1 + \dfrac{2}{x} + \dfrac{3}{x^2}$.

(2) If $p = 3\frac{1}{2}$, $q = \frac{3}{14}$, $r = 1\frac{2}{7}$, $s = 2\frac{1}{4}$ find the value of,
  (a) $p + q - r$,
  (b) pqs,
  (c) $\dfrac{ps}{r}$.

(3) If $A = 3\frac{1}{2}$, $B = 3\frac{3}{4}$, $C = 1\frac{1}{2}$, $D = 1\frac{1}{4}$, find the value of,
  (a) ACD,
  (b) $1 + AD$,
  (c) $\dfrac{CD}{AB}$.

(4) If $P = 3\frac{1}{3}$, $Q = 2\frac{1}{4}$, $R = 5\frac{3}{10}$, $S = 1\frac{1}{5}$, find the value of,
  (a) $P - Q$,
  (b) $R + S$,
  (c) $\dfrac{P - Q}{R + S}$.

(5) If $w = 1\frac{3}{7}$, $x = 1\frac{1}{5}$, $y = 1\frac{7}{9}$, $z = \frac{3}{4}$, find the value of,
  (a) $1 - wx$,
  (b) $1 + yz$,
  (c) $\dfrac{1 - wx}{1 + yz}$.

20 Scale Drawing

Before You Start

Always make sure that you have all the instruments which you will need. Do not waste your own time, and other people's, by borrowing, especially in an examination. You will need a sharp pencil, a ruler, compasses, set square, protractor, and a clean rubber. Many good drawings have been ruined by using a dirty rubber! This applies to all work with drawings of any kind.

Scale

The first thing to decide is the scale of your drawing. You may, of course, be given the scale. If not, make sure that the scale you choose will allow you to get the whole drawing on your paper. At the same time, it should take up as much of the paper as possible, and not be squashed into a corner, leaving most of the paper blank.

Always state clearly the scale of your drawing. When drawing maps, the ratio of any distance on the map to the corresponding distance on the ground, is called the *representative fraction*, or RF of the map.

Area

Remember that if the scale of a drawing is doubled, its area is increased four times. If the scale is trebled, the area is increased nine times.

Exercise 20a

(1) The plan of a house is drawn on a scale of 1 cm to 1 metre.
   (a) If the dining room is 3.5 m long, what is its length on the plan?

(b) If the area of the dining room is 10 m^2, what is its area on the plan?

(2) An athletics ground is drawn on a scale of 1 cm to 2 m. What is the length of the 100 m track as it appears on the drawing?

(3) The plan of a garden is drawn on a scale of 2 cm to 1 m.

(a) What is the true length of the garden path if its length on the plan is 30 cm?

(b) A flower bed has an area of 11 m^2. What is its area on the plan?

(4) A football stadium having an area of 4 hectares, is shown on a plan using a scale of 1 cm to 2 m. What is the area of the drawing of the stadium?

(5) Plans of a house are drawn on a scale of 5 cm to 1 m. What fraction of the ground area is covered by the plan?

(6) A map is drawn on a scale of 1 cm to 1 km. What is the representative fraction (RF) of the map?

(7) A map is drawn with an RF of $\frac{1}{100\,000}$. What is the distance between two towns, if the distance on the map is 5.5 cm?

(8) The map of a park is drawn with a RF of $\frac{1}{1\,000}$. If the area of the park is 150 ha, what is the area of the map?

(9) A rectangular room 20 m long and 15 m wide is to be laid with tiles which are 10 cm long and 7.5 cm wide. How many tiles will be needed?

(10) On a map with RF $\frac{1}{100\,000}$, what area is represented by,

(a) 1 cm^2, (b) 1 mm^2?

Exercise 20b

(1) The figure shows the plan of a garden, not drawn to scale. The path is 1 m wide.

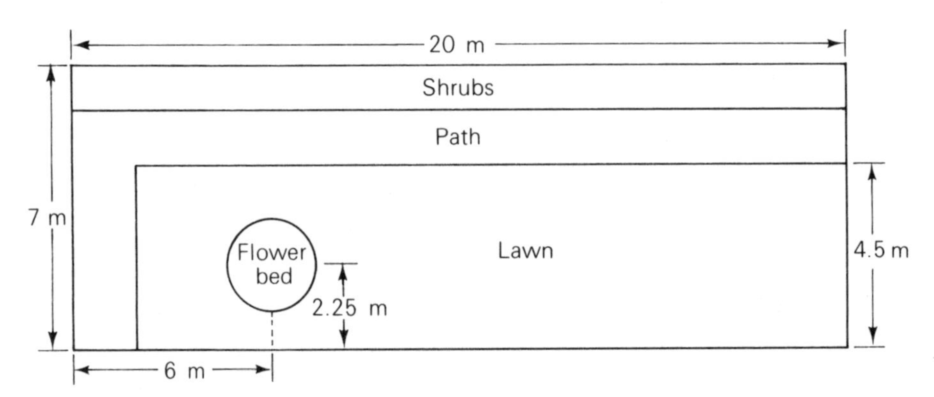

(a) Calculate the area of the flower bed, which is 3.5 m in diameter. Use the formula $A = \frac{22r^2}{7}$, and give your answer in m², correct to one decimal place.

(b) Calculate the area of the lawn in m², correct to one decimal place.

(c) Draw a plan of the garden, using a scale of 1 cm to 1 m.

(2) The figure shows the plan of a running track with semi-circular ends. It is not drawn to scale.

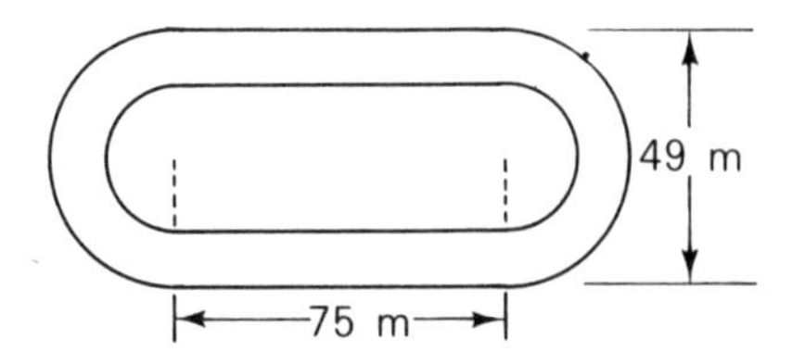

(a) If a plan of the track is drawn on a scale of 1 cm to 2 m, what is the overall length of the plan?

(b) A man runs once round the outside of the track. Does he run further along the straight edges, or along the curved? By how much is it greater? (The circumference of a circle is given by $C = \frac{44r}{7}$.)

(c) Make a plan of the track using a scale of 1 cm to 10 m.

(3) The figure shows the layout of a rectangular building plot, not drawn to scale.

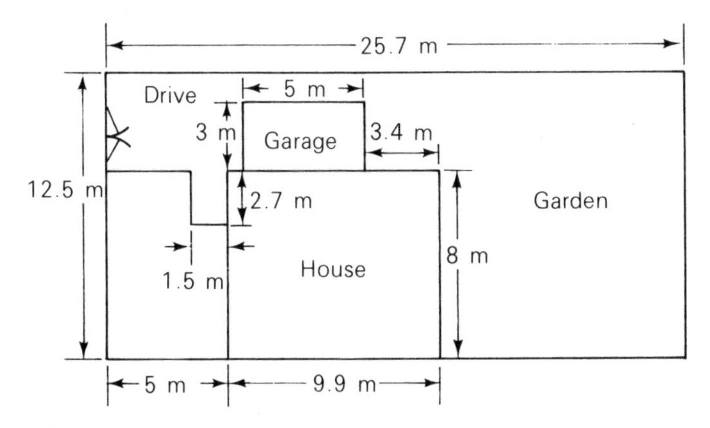

(a) How wide is the path at the side of the garage?

(b) What is the total ground area covered by the house and garage?

(c) Draw a scale drawing of the building plot, using a scale of 1 cm to 1 m.

(4) The figure shows the plan of a bungalow, not drawn to scale.

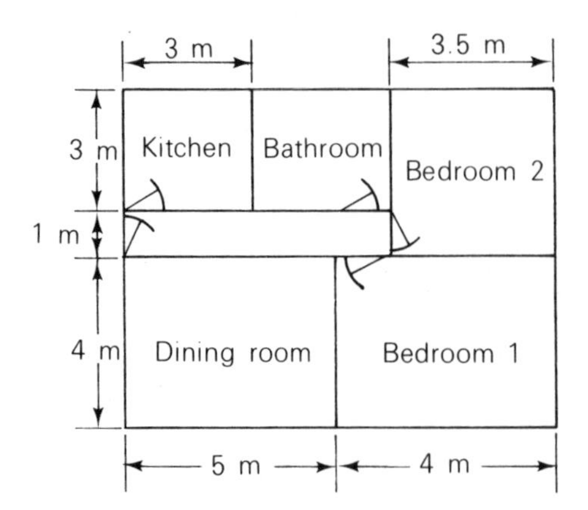

(a) What is the total ground area of the bungalow?
(b) What is the area of the bathroom floor?
(c) What will be the length of the dining room on a plan drawn to a scale of 4 cm to 1 m?
(d) Draw a plan of the bungalow using a scale of 2 cm to 1 m.

(5) The figure shows the plan of the ground floor of a cottage.

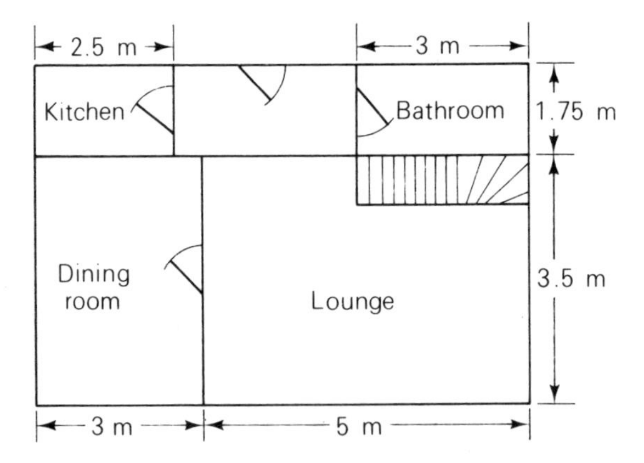

(a) If the ground floor originally consisted of lounge and dining room only, find the ground area of the original cottage.
(b) By what fraction was the ground area of the cottage increased, when the extension was built?
(c) Draw a plan of the complete ground floor using a scale of 4 cm to 1 m.

Section D
21 Number Systems

Number Bases

Usually we count in tens, and this is called the *denary* system. We say that we are counting in base 10. In this chapter we will look at other number bases and we will have to be sure which base we are working in. If we write $53_6 = 201_4$ this tells us that the number which is written as 53 (five-three) in base 6 is the same number that is written as 201 (two-zero-one) in base 4. The little 6 tells us that 53 is written in base 6 and the little 4 tells us that 201 is written in base 4. We can easily check that $53_6 = 201_4$ by changing both numbers to base 10.

Changing to Base 10

We can change 53_6 to base 10 by thinking of what each column of figures means. Reading from right to left, the column headings for base 10 are 1, 10, 10^2, 10^3, ... (1, 10, 100, 1 000, ...). In base 4 the column headings are 1, 4, 4^2, 4^3, ... (1, 4, 16, 64, ...).

Base 6

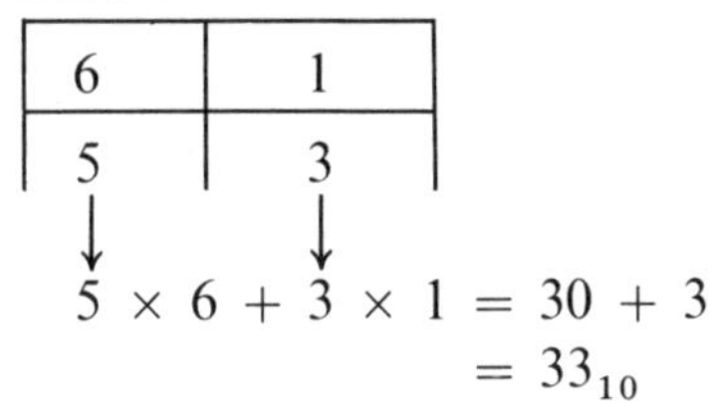

6	1
5	3

$$5 \times 6 + 3 \times 1 = 30 + 3 = 33_{10}$$

Base 4

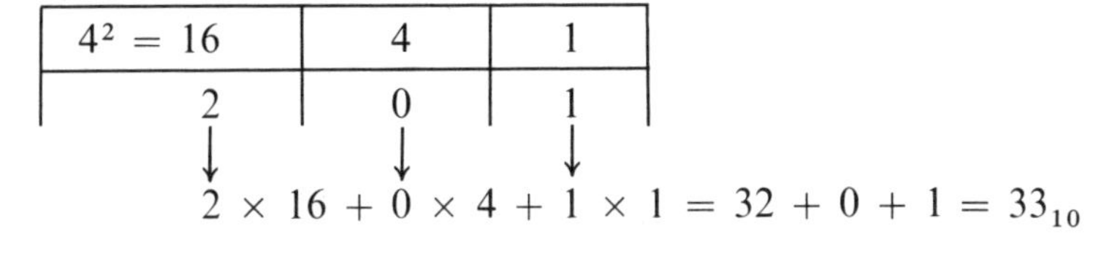

$4^2 = 16$	4	1
2	0	1

$$2 \times 16 + 0 \times 4 + 1 \times 1 = 32 + 0 + 1 = 33_{10}$$

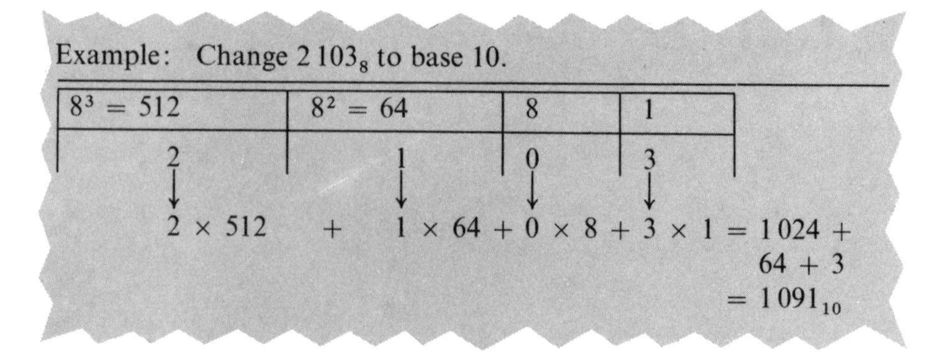

Example: Change $2\,103_8$ to base 10.

$8^3 = 512$	$8^2 = 64$	8	1
2	1	0	3

$$2 \times 512 + 1 \times 64 + 0 \times 8 + 3 \times 1 = 1024 + 64 + 3 = 1\,091_{10}$$

Changing from Base 10

To change $1\,091_{10}$ to base 8, we must first see which is the highest power of 8 which is smaller than 1 091. $8^3 = 512 < 1\,091$, but $8^4 = 4\,096 > 1\,091$. We must find how many 512's there are in 1 091, then how many 64's in the remainder and so on.

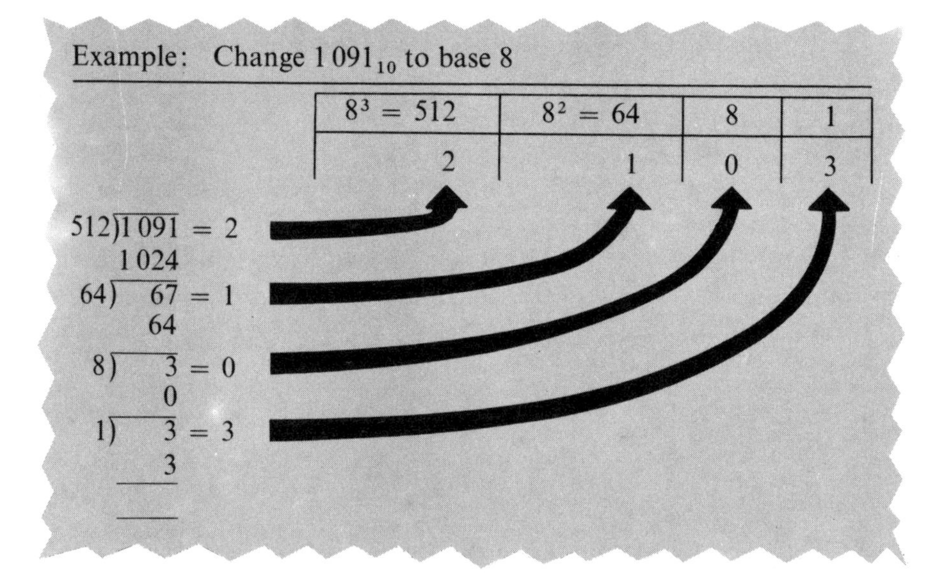

Example: Change $1\,091_{10}$ to base 8

$8^3 = 512$	$8^2 = 64$	8	1
2	1	0	3

512)1 091 = 2
1 024
64) 67 = 1
64
8) 3 = 0
0
1) 3 = 3
3

Base 12

So far, we have thought only of number bases less than 10. We could just as easily count in twelves or any other number greater than 10. As before the base number is shown by one-zero.

$12^2 = 144$	12	units
	1	0

If $10_{12} = 12_{10}$, what numeral in base 12 represents 10_{10}? We need a new numeral for this. Let us put $10_{10} = T_{12}$ and $11_{10} = E_{12}$.

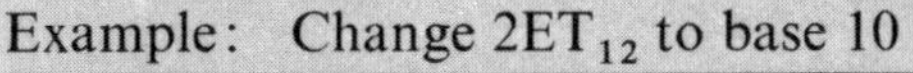

Example: Change $2ET_{12}$ to base 10

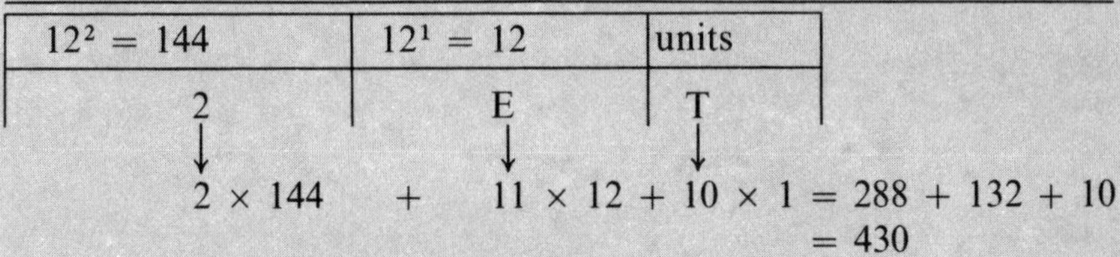

$12^2 = 144$	$12^1 = 12$	units
2	E	T

$$2 \times 144 + 11 \times 12 + 10 \times 1 = 288 + 132 + 10 = 430$$

$\therefore 2ET_{12} = 430_{10}$

Operations

When calculating with numbers in bases other than 10, you must be sure to remember which base you are working in.

Example: Calculate $253_6 + 105_6$. Check your answers by changing to base 10.

base 6		base 10
253	⟶	105
+ 105	⟶	41 +
402	⟶	146

When we work in base 10, shifting a number one place to the left multiplies it by 10_{10}. If we work in base 8, shifting a number one place to the left multiplies it by 10_8. That is, it multiplies it by eight. Whatever base we work in, shifting a number one place to the left multiplies it by the base number.

When a number is shifted one place to the left it is multiplied by the base number.

Example: Calculate (i) $(3\,562 \times 10)_6$ (ii) $(4\,735 \times 100)_8$

(i) $(3\,562 \times 10)_6 = 35\,620_6$ by shifting the numbers one place left.

(ii) $(4\,735 \times 100)_8 \doteq 473\,500_8$ by shifting the numbers two places left.

When multiplying numbers in bases other than 10, it will help if you first make a multiplication table for the base.

Example: Calculate $34_6 \times 12_6$

Base 6

×	1	2	3	4	5
1	1	2	3	4	5
2	2	4	10	12	14
3	3	10	13	20	23
4	4	12	20	24	32
5	5	14	23	32	41

Base 6

34 × 12
12
112
340
452

Exercise 21a

Change the following to base 10.

(1) 141_5
(2) 332_5
(3) 203_5
(4) $1\,101_5$
(5) $2\,301_5$
(6) 111_8
(7) 106_8
(8) 217_8
(9) 252_8
(10) 306_8
(11) 101_2
(12) 110_2
(13) $1\,011_2$
(14) $11\,011_2$
(15) $110\,100_2$
(16) 106_{12}
(17) $1E3_{12}$
(18) $20T_{12}$
(19) $21E_{12}$
(20) $2TE_{12}$
(21) 112_6
(22) 205_6
(23) $1\,031_6$
(24) $1\,324_6$
(25) $2\,001_6$

Change to base 5.

(26) 93_{10} (27) 121_{10} (28) 160_{10} (29) 380_{10} (30) 631_{10}

Change to base 2.

(31) 15_{10} (32) 19_{10} (33) 37_{10} (34) 60_{10} (35) 75_{10}

Change to base 3.

(36) 73_{10} (37) 92_{10} (38) 144_{10} (39) 162_{10} (40) 158_{10}

Change to base 7.

(41) 33_{10} (42) 42_{10} (43) 57_{10} (44) 68_{10} (45) 100_{10}

Change to base 12.

(46) 63_{10} (47) 87_{10} (48) 150_{10} (49) 290_{10} (50) 302_{10}

Calculate the following and check your answer by changing to base 10.

(51) $213_6 + 552_6$
(52) $456_8 + 241_8$
(53) $432_5 - 141_5$
(54) $11\,011_2 - 101_2$
(55) $26_7 \times 14_7$
(56) $23_5 \times 12_5$
(57) $421_5 \times 100_5$
(58) $416_7 \times 10_7$
(59) $11\,010_2 \div 10_2$
(60) $46T00_{12} \div 100_{12}$

Exercise 21b

(1) (a) Copy and complete the addition table for base 5.

+	1	2	3	4
1	2		4	
2		4		11
3		10	11	
4				13

(b) Change 531_{10} to base 5.

(c) Calculate $243_5 + 104_5$ and check your answer by changing to base 10.

(d) Calculate $3\,421_5 \times 10_5$.

(2) (a) Copy and complete the addition table for base 8.

+	1	2	3	4	5	6	7
1			4				
2		4					11
3						11	
4				10			
5				11	12		
6						14	
7		11					

(b) Change 612_{10} to base 8.

(c) Calculate $5\,670_8 \div 10_8$.

(d) Calculate $437_8 + 523_8$ and check your answer by changing to base 10.

(3) (a) Copy and complete the multiplication table for base 4.

×	1	2	3
1	1		
2		10	
3		12	

(b) Change 371_{10} to base 4.

(c) Calculate $33_4 \times 1\,000_4$.

(d) Calculate $23_4 \times 21_4$ and check your answer by changing to base 10.

(4) (a) Make a multiplication table for base 6.

(b) Change 279_{10} to base 6.

(c) Calculate $3\,520_6 \div 10_6$.

(d) Calculate $423_6 \times 4_6$ and check your answer by changing to base 10.

(5) (a) Change 636_{10} to base 8.

(b) Change $4E3_{12}$ to base 10.

(c) Calculate $11\,011_2 + 1\,001_2$.

(d) Calculate $221_3 - 122_3$.

(e) Calculate $243_7 \times 5_7$.

22 Percentages

Hundredths

Percentages are fractions written in hundredths.

$$\frac{35}{100} = 35\% \text{ (thirty-five per cent),}$$

$$\frac{93}{100} = 93\% \text{ (ninety-three per cent).}$$

Percentages are useful because they enable us to compare fractions. For example, it is not easy to see whether $\frac{2}{5}$ or $\frac{3}{8}$ is the larger fraction. If we change these fractions to percentages, we see that $\frac{2}{5} = \frac{40}{100} = 40\%$, and that $\frac{3}{8} = \frac{37.5}{100} = 37\frac{1}{2}\%$. It is now easy to see which is the larger.

Fractions and Decimals

To change a fraction or a decimal to a percentage, is to change it to hundredths.

$$\frac{3}{4} = \frac{75}{100} = 75\%,$$

$$0.42 = \frac{42}{100} = 42\%.$$

This can be done more simply by multiplying by 100.

Example: Change (i) $\frac{3}{4}$ (ii) 0.42 to percentages.

(i) $\frac{3}{\cancel{4}_1} \times \frac{\cancel{100}^{25}}{1} = 75\%$

(ii) $0.42 \times 100 = 42\%$

If a percentage is needed as a fraction or a decimal, divide by 100.

Example: Change 60% to (i) a fraction (ii) a decimal.

(i) $60\% = \frac{60}{100} = \frac{6}{10} = \frac{3}{5}$

(ii) $60\% = \frac{60}{100} = 0.6$

Remember, that to change a fraction to a decimal, you should divide the top number (numerator) by the bottom number (denominator).

Example: Change $\frac{3}{8}$ to a decimal.

```
  0.375
8)3.0
  24
   60
   56
    40
    40
    --
```

$\frac{3}{8} = 0.375$

Using Percentages

To find a percentage of an amount, remember that a percentage is a fraction in hundredths.

Example: Find 35% of £5.40

35% is $\frac{35}{100}$.

$\frac{35}{100}$ of £5.40 is £$\frac{35}{100} \times 5.40$

$= £\frac{35}{100} \times \frac{54}{10}$

$= £1.89$

```
35 × 54
  54
 140
1750
1890
```

Another way in which percentages are often used is in finding one number as a percentage of another. To do this, it must first be found as a fraction of the other.

Example: What percentage of 1 500 is 480?

480 is $\frac{480}{1\,500}$ of 1 500

$$\frac{480}{1\,500} = \frac{48}{150} = \frac{8}{25} = 0.32 = 32\%.$$

$$\begin{array}{r} 0.32 \\ 25\overline{)8.0} \\ 75 \\ \hline 50 \\ 50 \\ \hline \end{array}$$

(This can be done more simply by saying $\frac{8}{25} = \frac{32}{100} = 32\%$.)

Exercise 22a

Write as percentages

(1) $\frac{43}{100}$ (2) $\frac{22}{100}$ (3) $\frac{4}{100}$ (4) $\frac{123}{100}$ (5) $2\frac{3}{100}$ (6) $\frac{4}{10}$ (7) $\frac{3}{10}$ (8) $\frac{7}{20}$ (9) $\frac{3}{50}$ (10) $\frac{7}{25}$ (11) $\frac{3}{40}$ (12) $\frac{3}{8}$ (13) $\frac{5}{16}$ (14) $2\frac{4}{5}$ (15) $2\frac{21}{60}$ (16) 0.75 (17) 0.35 (18) 0.01 (19) 2.67 (20) 5.43

Write as decimals

(21) 35% (22) 21% (23) 43% (24) 121% (25) 235% (26) $\frac{2}{5}$ (27) $\frac{3}{4}$ (28) $\frac{5}{8}$ (29) $\frac{3}{8}$ (30) $\frac{3}{5}$

Write as decimals correct to 2 decimal places

(31) $\frac{3}{7}$ (32) $\frac{5}{6}$ (33) $\frac{7}{11}$ (34) $\frac{4}{9}$ (35) $\frac{2}{3}$

Write as decimals correct to 3 significant figures

(36) $\frac{5}{7}$ (37) $\frac{2}{3}$ (38) $\frac{1}{6}$ (39) $\frac{3}{11}$ (40) $\frac{7}{9}$

Calculate the following

(41) 30% of 560 (42) 40% of 1 year (43) 32% of £43 (44) 43% of 21 km (45) 35% of 360° (46) 73% of 500 m^2

(47) 120% of 750 ml
(48) 15% of 2 hours
(49) 8% of 37.5 m^3
(50) 35% of 360°

(51) What percentage of 5 000 is 450?
(52) What percentage of 450 is 108?
(53) What percentage of £2.20 is 77p?
(54) What percentage of 325 km is 91 km?
(55) What percentage of 17.5 l is 5.6 l?
(56) What percentage of 20 weeks is 21 days?
(57) What percentage of 360° is 72°?
(58) What percentage of £20 is £4.60?
(59) What percentage of 250 cm^3 is 70 cm^3?
(60) What percentage of 45 m^2 is 27 m^2?

Exercise 22b

(1) A garden which is 40 m long and 20 m wide contains a lawn which is 30 m long and 9 m wide. Calculate:
(a) the length of the lawn as a percentage of the length of the garden
(b) the width of the lawn as a percentage of the width of the garden
(c) the area of the lawn as a percentage of the area of the garden
(d) the cost of turfing the lawn is £67.50, calculate the cost of fertiliser, if this is 8% of the cost of turf.

(2) At Hilltop school, there are 424 boys and 376 girls. 256 pupils go home to lunch each day and the rest have school lunch. Of those who go home to lunch 37.5% are boys. Calculate:
(a) the percentage of boys in the school
(b) the number of boys who go home to lunch
(c) the number of girls who go home to lunch as a percentage of all the pupils in the school
(d) the percentage attendance on a day when there are 44 pupils absent.

(3) A man buys a house for £13 600. The rateable value of the house is £340. Calculate:
(a) the general rate charged at 67% of the rateable value
(b) the water rate, if this is 5% of the general rate
(c) the rateable value, as a percentage of the price of the house
(d) the percentage profit, if the house is sold for £16 320.

(4) The amounts of tea from the tea-growing regions of the world are given in the table below, in thousands of tonnes. The percentages of total world production are also given.

region	thousands of tonnes	%
India		32
Ceylon	250	
China		
Others	450	36
Total	1 250	100

(a) Calculate the number of tonnes grown in India.
(b) Calculate the percentage for Ceylon.
(c) Calculate the number of tonnes grown in China.
(d) Calculate the percentage for China.
(e) Show your percentages on a pie chart.

(5) The table shows the population of some countries in South America.

Country	population in millions	%
Argentina	21	
Brazil		
Colombia	18	12
Peru		8
Others	21	14
Total	150	100

(a) Calculate the population of Peru.
(b) Calculate the percentage for Argentina.
(c) Calculate the population of Brazil.
(d) Calculate the percentage for Brazil.
(e) Show your percentages on a pie chart.

23 Four-Figure Logarithms

Logarithms

Part of the logarithm tables is shown below. The first column, shown by the number 1 arrow gives all numbers from 10 to 99. The second column, shown by the number 2 arrow, gives logarithms correct to 4 significant figures. These two columns are used in the same way as the 3-figure tables which you have already used. Follow the programme to find how to use the tables for 4-figure numbers. You will need to look at the table below, and to have your own log table in front of you.

LOGARITHMS OF NUMBERS

	0	1	2	3	4	5	6	7	8	9	Differences 1	2	3	4	5	6	7	8	9
10	0000	0043	0086	0128	0170	0212	0253	0294	0334	0374	4	8	12	17	21	25	29	33	37
11	0414	0453	0492	0531	0569	0607	0645	0682	0719	0755	4	8	11	15	19	23	26	30	34
12	0792	0828	0864	0899	0934	0969	1004	1038	1072	1106	3	7	10	14	17	21	24	28	31
13	1139	1173	1206	1239	1271	1303	1335	1367	1399	1430	3	6	10	13	16	19	23	26	29
14	1461	1492	1523	1553	1584	1614	1644	1673	1703	1732	3	6	9	12	15	18	21	24	27
15	1761	1790	1818	1847	1875	1903	1931	1959	1987	2014	3	6	8	11	14	17	20	22	25
16	2041	2068	2095	2122	2148	2175	2201	2227	2253	2279	3	5	8	11	13	16	18	21	24
17	2304	2330	2355	2380	2405	2430	2455	2480	2504	2529	2	5	7	10	12	15	17	20	22
18	2553	2577	2601	2625	2648	2672	2695	2718	2742	2765	2	5	7	9	12	14	16	19	21
19	2788	2810	2833	2856	2878	2900	2923	2945	2967	⊕	2	4	7	9	11	13	16	18	20
20	3010	3032	3054	3075	3096	3118	3139	3160	3181	3201	2	4	6	8	11	13	15	17	19

Answers ↓	
	5 300 = 5.3 × 10^3. The characteristic of log 5 300 is 3. 120 = 1.2 × 10^2. What is the characteristic of log 120?
2	Look at page 106 and follow the number 3 arrow with the first finger of your left hand, until you get to 12 in the first column. The decimal part of log 12 has a ring round it. What is its value?
0792	This means that log 120 is 2.0792. We have used the same method as we would do with 3-figure tables. Look at page 106. What is the value of log 17?
1.2304	Now use your own log table. What is log 73?
1.8633	What is log 870?
2.9395	Now to find log 1 550. What is the characteristic of log 1 550?
3	What are the first two figures of 1 550?
15	Look at the logarithms on page 106. Move your finger down the first column until you get to 15. Then follow the number 4 arrow across the page until you get to the 5 column. What number do you get to?
1903	The characteristic of log 1 550 is 3 and the number in the tables is 1 903, so log 1 550 = 3.1903. Look at page 106. What is the value of log 1.78?
0.2504	We will now find the logarithm of a 4-figure number. What is the characteristic of log 19.73?
1	Look at the logarithm table on page 106. Move your first finger down the first column until you get to 19, then follow the number 5 arrow across the page to the 7 column. What number do you find?
2 945	What is the value of log 19.7?
1.2945	We know that log 19.7 = 1.2945, and log 19.73 is a little bit more. We can find how much more from the difference columns. Put the first finger of your left hand on 2 945 in the tables and move the first finger of your right hand across the page until you get to the 3 difference column. What number do you find?

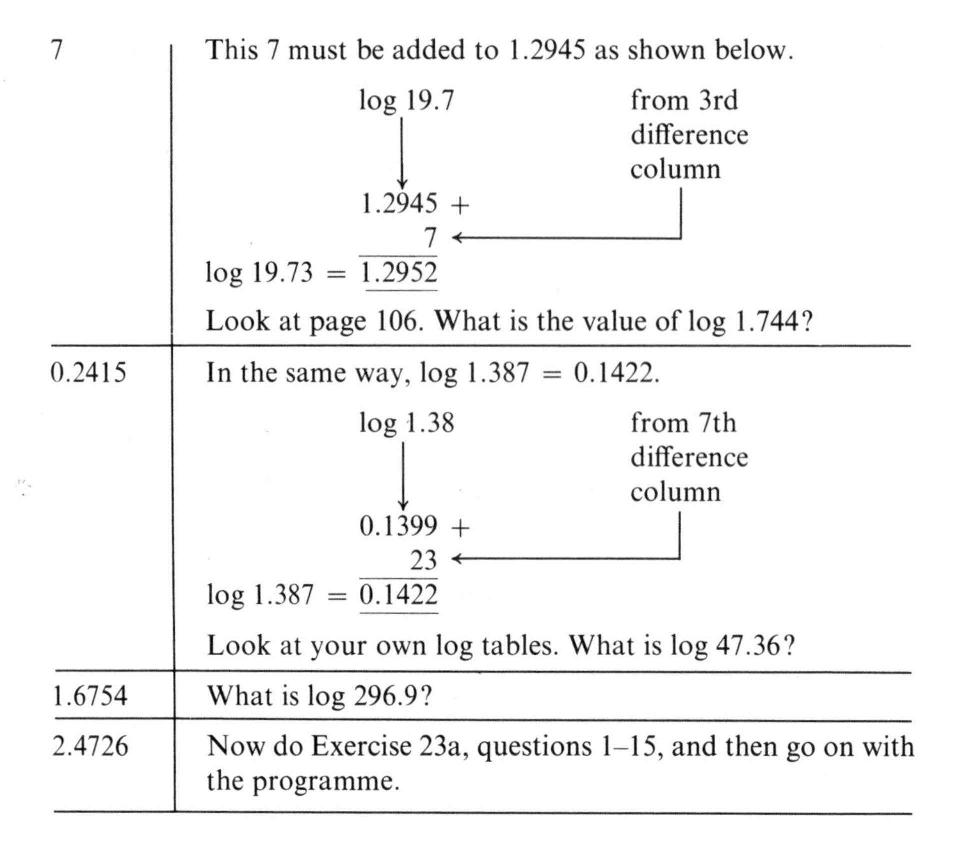

7	This 7 must be added to 1.2945 as shown below. log 19.7 → 1.2945 + 7 (from 3rd difference column) log 19.73 = 1.2952 Look at page 106. What is the value of log 1.744?
0.2415	In the same way, log 1.387 = 0.1422. log 1.38 → 0.1399 + 23 (from 7th difference column) log 1.387 = 0.1422 Look at your own log tables. What is log 47.36?
1.6754	What is log 296.9?
2.4726	Now do Exercise 23a, questions 1–15, and then go on with the programme.

Anti-Logarithms

Anti-logarithm tables are used for going back the other way. To find the number whose logarithm is 2.0747 we look up 0747 in the anti-log tables. The programme will show you how to do this.

Answers ↓	
	What are the first two figures of 0747?
07	Look at the anti-log table on page 109. Move your finger down the first column until you get to .07. Keep your finger on .07. What is the 3rd figure of 0747?
4	Follow the arrow across the page with your finger until you get to the 4 column. What number do you find?

1 186	Keep your finger on 1 186. What is the 4th figure of 0747?
7	Keep the first finger of your left hand on 1 186 and move the first finger of your right hand across the page to the 7 difference column. What number do you find?
2	This 2 must be added to 1 186, and we get 1 188. This means that log 1.188 is 0.0747. What is the number whose log is 2.0747?
118.8	Look at page 109. What is the anti-log of 1.1326?
13.57	Look at your own anti-log tables. What is the anti-log of 0.6258?
4.225	What is the anti-log of 3.4374, correct to 3 significant figures?
2 740	Now do Exercise 23a, questions 16–30, and then go on to page 110.

ANTILOGARITHMS

	0	1	2	3	4	5	6	7	8	9	Differences								
											1	2	3	4	5	6	7	8	9
.00	1000	1002	1005	1007	1009	1012	1014	1016	1019	1021	0	0	1	1	1	1	2	2	2
.01	1023	1026	1028	1030	1033	1035	1038	1040	1042	1045	0	0	1	1	1	1	2	2	2
.02	1047	1050	1052	1054	1057	1059	1062	1064	1067	1069	0	0	1	1	1	1	2	2	2
.03	1072	1074	1076	1079	1081	1084	1086	1089	1091	1094	0	0	1	1	1	1	2	2	2
.04	1096	1099	1102	1104	1107	1109	1112	1114	1117	1119	0	1	1	1	1	2	2	2	2
.05	1122	1125	1127	1130	1132	1135	1138	1140	1143	1146	0	1	1	1	1	2	2	2	2
.06	1148	1151	1153	1156	1159	1161	1164	1167	1169	1172	0	1	1	1	1	2	2	2	2
.07	1175	1178	1180	1183	1186	1189	1191	1194	1197	1199	0	1	1	1	1	2	2	2	2
.08	1202	1205	1208	1211	1213	1216	1219	1222	1225	1227	0	1	1	1	1	2	2	2	3
.09	1230	1233	1236	1239	1242	1245	1247	1250	1253	1256	0	1	1	1	1	2	2	2	3
.10	1259	1262	1265	1268	1271	1274	1276	1279	1282	1285	0	1	1	1	1	2	2	2	3
.11	1288	1291	1294	1297	1300	1303	1306	1309	1312	1315	0	1	1	1	2	2	2	2	3
.12	1318	1321	1324	1327	1330	1334	1337	1340	1343	1346	0	1	1	1	2	2	2	2	3
.13	1349	1352	1355	1358	1361	1365	1368	1371	1374	1377	0	1	1	1	2	2	2	3	3
.14	1380	1384	1387	1390	1393	1396	1400	1403	1406	1409	0	1	1	1	2	2	2	3	3
.15	1413	1416	1419	1422	1426	1429	1432	1435	1439	1442	0	1	1	1	2	2	2	3	3

Using Logarithms

If two or more numbers are to be multiplied, their logarithms should be added. For division, the logarithms must be subtracted. The fourth figure on the tables is not accurate, so your answers should always be rounded off to 3 significant figures.

Example: Calculate (i) 3.742 × 41.68 × 0.03128
(ii) 91.85 ÷ 34.67.

Give both your answers correct to 3 significant figures.

(i)

	No	Log	
×	3.742	0.5731	+
	41.68	1.6199	
	0.03128	$\bar{2}.4953$	
	4.878	0.6883	

3.742 × 41.68 × 0.03128
= 4.88 (correct to 3 sig. figs.).

(ii)

	No	Log	
÷	91.85	1.9630	−
	34.67	1.5400	
	2.649	0.4230	

91.85 ÷ 34.67 = 2.65
(correct to 3 sig. figs.).

Exercise 23a

Use your tables to find the logarithms of:

(1) 53.72
(2) 41.85
(3) 7.321
(4) 4.906
(5) 371.2
(6) 471.3
(7) 849.5
(8) 9 402
(9) 7 430
(10) 249
(11) 0.6213
(12) 0.4142
(13) 0.03161
(14) 0.4112
(15) 0.003214

Use your anti-log tables to find the number whose logarithm is:

(16) 2.7131
(17) 1.8114
(18) 3.0421
(19) 2.1163
(20) 0.4185
(21) 0.9472
(22) 0.8321
(23) 4.6739
(24) 2.3071
(25) 1.8164
(26) $\bar{1}.6213$
(27) $\bar{1}.8144$
(28) $\bar{2}.8731$
(29) $\bar{3}.4118$
(30) $\bar{1}.6511$

Calculate the following correct to 3 significant figures

(31) 5.731 × 21.42
(32) 74.01 × 2.335
(33) 6.71 × 42.62
(34) 543.2 × 0.5621
(35) $(11.21)^2$
(36) $(5.963)^2$
(37) 2.041 × 5.763 × 24.12
(38) 8.743 × 763.1 × 0.4268

(39) $563.7 \times 26.14 \times 0.03211$
(40) $456.6 \times 974.8 \times 0.003485$
(41) $63.72 \div 4.855$
(42) $191.6 \div 7.332$
(43) $593.7 \div 21.25$
(44) $1\,408 \div 261.8$
(45) $34.2 \div 4.244$
(46) $68.93 \div 92.86$
(47) $43.14 \div 52.91$
(48) $2.76 \div 83.67$
(49) $0.6213 \div 2.842$
(50) $0.8341 \div 15.76$

Exercise 23b

(1) If $P = 37.32$ and $Q = 46.54$, use logarithms to calculate:
(a) $P \times Q$
(b) $Q \div P$
(c) $P \div Q$
(d) P^2
correct to 3 significant figures.
(2) If $a = 346.1$, $b = 0.4623$ and $c = 11.73$, use logarithms to calculate:
(a) ab
(b) $b \div a$
(c) b^2
(d) abc
correct to 3 significant figures.
(3) Calculate the following correct to 3 significant figures:
(a) 61.72×11.73
(b) $2.61 \times 21.83 \times 0.073$
(c) $596.6 \div 24.32$
(d) $(7.692)^2$.
(4) If $r = 2.327$ and $h = 14.55$ calculate
(a) $2\pi r$
(b) πr^2
(c) $2\pi rh$
(d) $\pi r^2 h$.
Take $\pi = 3.142$ and give your answers correct to 3 significant figures.

24 Simple Equations

Easy Equations

An equation which has an x, but no x^2 or higher powers of x is called a *simple equation*. There may be more than one x, or the unknown number may be shown by a different letter.

Example: Solve the equation $5p + 2 = 7p - 3$.

1. Write the equation. $5p + 2 = 7p - 3$
2. Subtract $5p$ from both sides. $2 = 2p - 3$
3. Add 3 to both sides. $5 = 2p$
4. Divide both sides by 2. $p = 2\frac{1}{2}$.

Equations with Brackets

First remove the brackets by multiplying.

Example: Solve the equation $5(4x + 1) = 7(2x - 1)$.

1. Write the equation. $5(4x + 1) = 7(2x - 1)$
2. Multiply $(4x + 1)$ by 5 and $(2x - 1)$ by 7. $20x + 5 = 14x - 7$
3. Subtract $14x$ from both sides. $6x + 5 = -7$
4. Subtract 5 from both sides. $6x = -12$
5. Divide both sides by 6. $x = -2$.

Equations with Fractions

To solve an equation such as

$$\frac{x+6}{2}+\frac{x-3}{5}=1$$

we must first get rid of the fractions. If we multiply each term in the equation by 10, the two sides will still be equal.

We choose 10 to multiply by because both 2 and 5 are factors of 10. We are then able to cancel both denominators and are left with whole numbers only.

Example: Solve the equation $\frac{x+6}{2}+\frac{x-3}{5}=1.$

1. Write the equation. What is the smallest multiple of 2 and 5?

$$\frac{x+6}{2}+\frac{x-3}{5}=1$$

2. The smallest multiple of 2 and 5 is 10. Multiply each term by 10.

$$\frac{10(x+6)}{2}+\frac{10(x-3)}{5}=10$$

3. Cancel.

$$5(x+6)+2(x-3)=10$$

4. Remove the brackets.

$$5x+30+2x-6=10$$

5. Collect like terms on the LHS.

$$7x+24=10$$

6. Subtract 24 from both sides.

$$7x=-14$$

7. Divide both sides by 7.

$$x=-2.$$

Solving Problems

Equations are used to solve problems. If you know that twice a number is 6, then you can find that the number is 3 because 3 is half of six. If we let the unknown number be x, then we know that $2x = 6$, so that $x = 3$.

To solve a problem, first state the given facts in mathematical language and then solve the resulting equation or equations.

Example: A man drives from Bristol to Bath, and then another 9 km to the village of Box. After visiting friends he drives back along the same road. When he arrives at Bristol he sees that he has driven a total of 52 km. What is the distance from Bristol to Bath?

Let the distance from Bristol to Bath be x km.
Then the distance from Bristol to Box is $(x + 9)$ km.
The distance from Bristol to Box and back is $2(x + 9)$ km.

$$\therefore 2(x + 9) = 52$$
$$x + 9 = 26$$
$$x = 17.$$

The distance from Bristol to Bath is 17 km.

Exercise 24a

Solve the following equations:

(1) $5x = 25$
(2) $3x = -27$
(3) $4y = 2$
(4) $2x = 3$
(5) $3x = -2$
(6) $3x + 2 = 11$
(7) $5x - 4 = 11$
(8) $4q + 5 = 7$
(9) $3x + 5 = 2$
(10) $8x + 3 = 1$
(11) $3x + 7 = 7x - 1$
(12) $2x - 1 = 3x - 6$
(13) $2x - 3 = 4x + 1$
(14) $3x + 4 = -3x$
(15) $4p = 6p + 2$
(16) $3(2x + 1) = 7$
(17) $5(3x - 2) = -7$
(18) $4(x + 1) = 5$
(19) $3(2 - x) = 4$
(20) $2(5x + 3) = 12$
(21) $2(3x + 1) = 7x - 2$
(22) $3(m - 2) = m + 6$
(23) $7(z + 1) = 10z + 1$
(24) $3(4x + 5) = 4x + 3$
(25) $2(3x - 1) = 9x + 1$
(26) $4(x - 1) = 3(x + 1)$
(27) $2(x + 2) = 7(x - 3)$
(28) $5(2x - 1) = 2(4x - 3)$
(29) $3(x + 2) = 7(x - 2)$
(30) $4(3x + 1) = 5(6x - 1)$
(31) $\frac{x - 3}{2} = \frac{x + 1}{6}$
(32) $\frac{x + 3}{2} = \frac{5x - 3}{4}$
(33) $\frac{7x - 1}{6} = \frac{2x + 1}{3}$
(34) $\frac{x + 3}{7} = \frac{x - 5}{3}$
(35) $\frac{x + 4}{2} = \frac{2x + 3}{3}$
(36) $\frac{x + 5}{2} + \frac{x + 4}{3} = 3$
(37) $\frac{4x + 3}{5} - \frac{x}{3} = 2$
(38) $\frac{x + 5}{3} - \frac{x}{2} = 2$
(39) $\frac{2x + 3}{2} + \frac{4x + 1}{3} = 3$
(40) $\frac{2x + 3}{3} + \frac{x}{2} = 1$

(41) I think of a number and double it. If I now add 3, the result is 7. What was the number I first thought of? (Let the number be x.)
(42) I think of a number and add 3. If I now double the sum, the result is 16. What was the number I first thought of? (Let the number be x.)
(43) Bob is 5 years older than Sally. The sum of their ages is 21 years. How old is Sally? (Let Sally's age be x years.)
(44) A rectangle is 4 cm wide. How long is the rectangle if the perimeter is 22 cm? (Let the length be x cm.)
(45) Anne and Joan go shopping. Anne buys a book for £2 and a coat. Joan spends the same amount of money as Anne. Between them, they spend £22. How much did Anne pay for the coat? (Let the cost of the coat be £x.)

Exercise 24b

(1) (a) Find the value of $2(3x + 2)$ if $x = 4$.
(b) Solve the equation $x + 5 = 2x + 1$.
(c) Solve the equation $3(x + 2) = 2(2x + 1)$.
(d) A man walks from his home to a shop and then walks another 1 km to a friend's house. He then goes back home the way he came, so that he walks a total of 6 km. How far is it from his house to the shop?
(2) (a) Find the value of $3(5x - 2)$ if $x = -1$.
(b) Solve the equation $5x + 3 = 10x - 1$.
(c) Solve the equation $\frac{x + 1}{2} + \frac{x}{3} = 3$.
(d) Ken is twice as old as Ted. In three years' time Ken will be 13 years old. Make up a simple equation from these facts and solve it to find Ted's age.
(3) (a) Solve the equation $6x + 3 = 4x + 7$.
(b) Solve the equation $3(x + 1) = 4(x - 1)$.
(c) Solve the equation $\frac{x + 2}{2} = \frac{x + 4}{3}$.
(d) I think of a number and subtract 2. If I now double the result, I get 10. What was the number I first thought of?
(4) (a) Find the value of $\frac{x}{3} - 2$ if $x = 6$.
(b) Solve the equation $4(x + 1) = 3(x + 2)$.
(c) Solve the equation $\frac{4x - 3}{5} + \frac{x}{2} = 2$.
(d) A purse contains 1p and 2p coins. There are three times as many 1p coins as there are 2p coins. If a 10p piece is added to the purse, the total value is £1. Make up an equation and solve it to find the number of 2p coins.

25 Locus and Envelope

Sets of Points

Let O be a fixed point and let P be a point on the paper which is 1 cm from O. There are many possible positions for P. If we mark all the possible positions for P, we would have an infinite set of points.

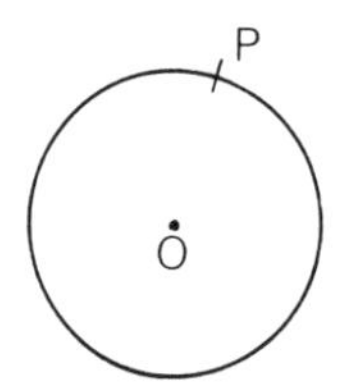

This set of points is called the *locus* of P. If P moves through all the possible positions, then the locus is the path traced out by P. In this case, the locus is a circle centre O, radius 1 cm.

Sets of Lines

Suppose a ladder leans against a wall and slips down. The positions of the ladder are shown below.

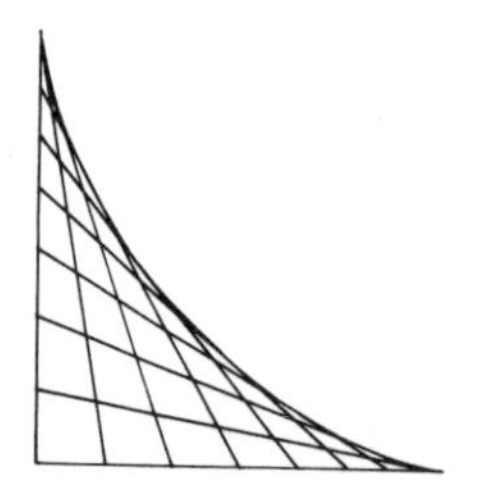

Even though the diagram is made up of straight lines, the result is a curve. If we draw a set of straight lines which result in a curve, then the curve is called the *envelope* of the lines.

If the pattern is repeated in all four quadrants as shown below, it makes a curve which is called an astroid.

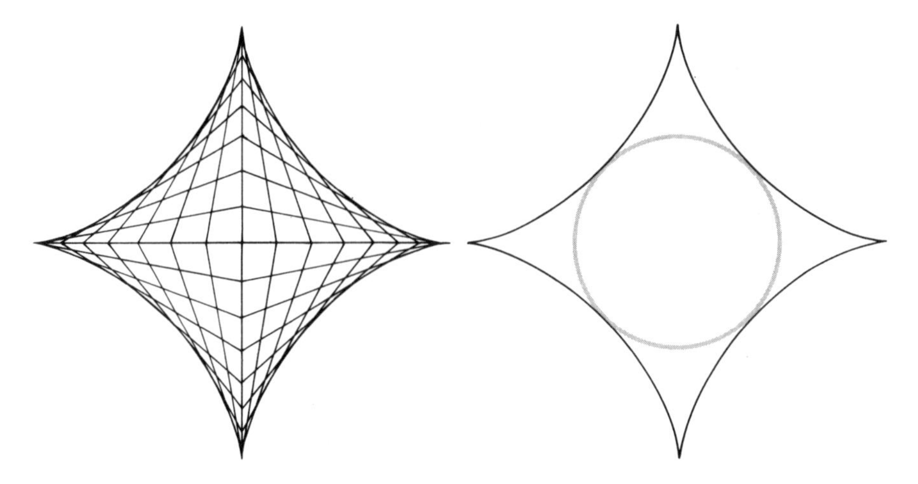

If we mark all the positions of the mid-point of the 'ladder', we find that the locus is a circle.

Exercise 25a

(1) Mark a point P in your exercise book and draw the locus of a point which moves on the paper so that it is always 4 cm from P. What kind of curve is it?
(2) In question 1, what would the locus be if the point were free to move off the paper?
(3) Draw a line PQ, 5 cm long. Draw the locus of a point X which moves so that XP = XQ. What line have you drawn?
(4) In question 3, what would the locus be if X is free to move off the paper?
(5) Draw a circle radius 5 cm. Draw a set of points on the paper which are all 1 cm from your circle. What have you drawn?
(6) Lay a piece of paper on a drawing board and put two drawing pins into the board, about 4 cm apart. Make a loop of thin string about 10 cm long and put it over the pins. Draw the string tight with the point of a pencil, and draw a curve as shown.

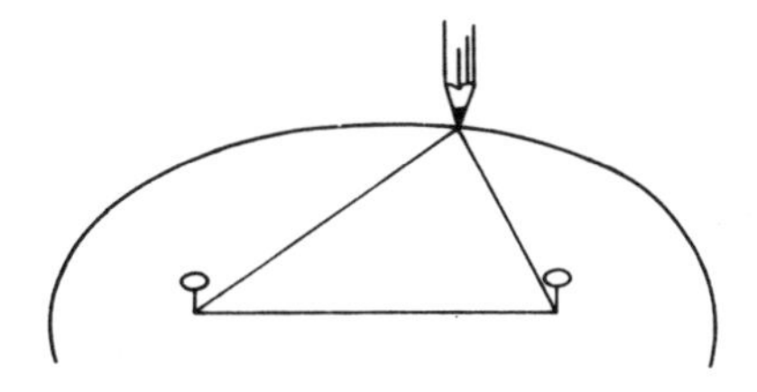

The path traced out by the tip of your pencil is called an *ellipse*. An ellipse is the curve traced out by the planets as they move round the sun.

(7) Describe the set of all points on your paper which are not more than 5 cm from a fixed point P.

(8) Describe the set of all points in space which are not more than a million km from the moon.

(9) Describe the locus of a point on the edge of a door as the door opens.

(10) Draw two lines, each 10 cm long at right angles. Let the vertical line represent a wall and let the horizontal line represent a floor. Draw lines 10 cm long on your figure to show the positions of a 'ladder' as it slides down the wall. Draw over the envelope of the lines in red. Mark the mid-point of each line and draw the locus of the mid-points.

Exercise 25b

(1) Draw a line AB, 6 cm long and construct the perpendicular bisector.

(a) Mark two points P and Q on the perpendicular bisector, each of which are 4 cm from AB.

(b) Measure AP, PB, BQ, QA. What kind of figure is APBQ?

(c) Describe the locus of points which are at equal distances from A and B.

(d) Describe the locus of points which are at equal distances from P and Q.

(2) Draw a line AB, 8 cm long and mark its mid-point X.

(a) Use your protractor to draw a line PXQ, so that $\angle QXB = 60°$.

(b) Without using your protractor, construct KL, the bisector of $\angle QXB$.

(c) Without using your protractor, construct MN, the bisector of $\angle PXB$.

(d) Describe the set of points which are at equal distances from AB and PQ.

(3) (a) Construct $\triangle ABC$ in which $AC = 8$ cm, $BC = 6$ cm and $\angle C = 90°$.

(b) Measure the hypotenuse of $\triangle ABC$ and check your answer by using the Pythagoras rule. Show your working.

(c) Draw the circumcircle of $\triangle ABC$.

(d) If P is any point so that $\angle APC$ is a right angle, describe the locus of P.

(4) Draw a circle of radius 5 cm and use your protractor to mark off the circumference of the circle into 36 equal parts.

(a) Number the marks on your circle from 1–36 and use the circle to show the mapping $n \to n + 10$.

(b) Describe the envelope of the lines.

(c) Is the envelope for the mapping $n \to n + 20$ larger or smaller than the one you have drawn?

(d) For which mapping does the envelope vanish?

(5) Describe the locus of

(a) the tip of the minute hand of a clock as the hand rotates for 1 hour,

(b) the centre of a bicycle wheel as the wheel moves forward,

(c) the bob of a pendulum as it swings backwards and forwards,

(d) the tip of the needle of a record player as the record plays.

26 Regular Polygons

Drawing Regular Polygons

A polygon is a plane figure with 3 or more sides. A regular polygon has all sides equal and all angles equal. A regular polygon is made up of isosceles triangles. A regular 5-sided polygon is made up of 5 isosceles triangles. A regular 8-sided polygon is made up of 8 isosceles triangles. All regular polygons fit into a circle.

If we talk about a regular polygon which could have any number of sides, we call it n-sided. A regular n-sided polygon is made up of n isosceles triangles.

For a regular 5-sided polygon, the circle is divided into 5 equal parts and this shows us how to draw a regular 5-sided polygon.

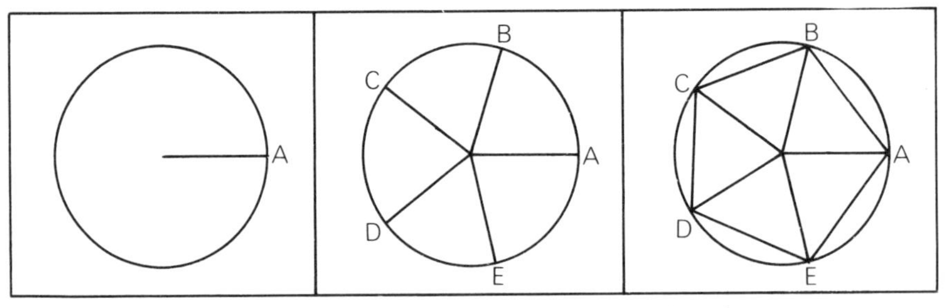

1. Draw a circle and any one radius

2. For any regular 5-sided polygon, each angle at the centre is $360° \div 5 = 72°$. Draw 5 angles, each 72°

3. Join AB, BC, CD, DE, EA.

Naming Polygons

A regular 3-sided polygon is an equilateral triangle. A regular 4-sided polygon is a square. Other regular polygons have names which say how many sides they have. The names of some polygons are given below.

5-sided .... *pentagon*
6-sided .... *hexagon*
8-sided .... *octagon*
10-sided .... *decagon*
12-sided .... *dodecagon*

Angles of a Regular Polygon

Each angle at the centre of a regular polygon is found by dividing 360° by the number of sides.

Each angle at the centre of a regular pentagon is $360° \div 5 = 72°$.
Each angle at the centre of a regular octagon is $360° \div 8 = 45°$.
Each angle at the centre of a regular n-sided polygon is $360° \div n$.

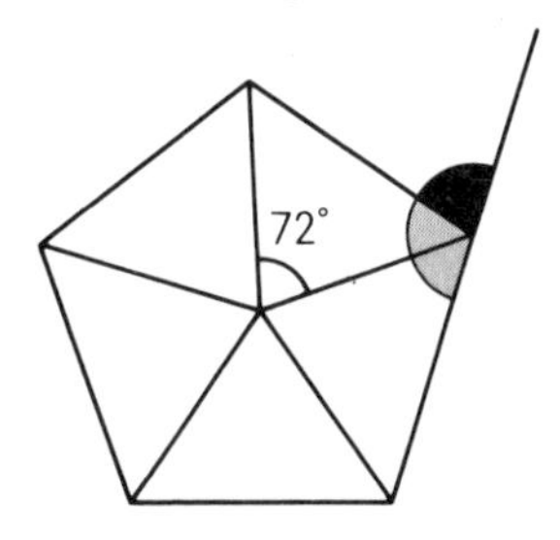

The black angle in the figure is called an *exterior angle* because it is outside the figure. It is equal to each angle at the centre, which in this case is 72°. Thus, the exterior angle of a regular n-sided polygon is $360° \div n$.

The exterior angle of a regular n-sided figure is $\frac{360°}{n}$.

The grey angle in the figure is called the *interior angle* because it is inside the figure. It can be found by subtracting the exterior angle from 180°. For a regular pentagon, the interior angle is $180° - 72° = 108°$.

If the exterior angle of a regular polygon is not known, the interior angle can still be calculated.

For a regular n-sided polygon, the exterior angle is $\frac{360°}{n} = \frac{4}{n}$ right angles. Thus, the interior angle is $180° - \frac{360°}{n}$ or $(2 - \frac{4}{n})$ right angles. There are n interior angles, so the sum of all the interior angles is $n(2 - \frac{4}{n})$ right angles $= (2n - 4)$ right angles.

The sum of the interior angles of a regular n-sided figure is $(2n - 4)$ right angles.

Exercise 26a

Calculate, in degrees, the sum of the interior angles for each of the following regular figures:

(1) hexagon (2) octagon (3) dodecagon (4) 17-sided.

Calculate the size of each angle at the centre of each of the following regular figures:

(5) octagon (6) decagon (7) 20-sided (8) 36-sided.

Calculate the size of an exterior angle for each of the following regular figures:

(9) 9-sided (10) 40-sided (11) 12-sided (12) 30-sided.

Calculate the size of an interior angle for each of the following regular figures:

(13) 8-sided (14) 6-sided (15) 12-sided (16) 5-sided.

Draw each of the following regular figures in a circle radius 5 cm and measure the length of the side of your figure, correct to 0.1 cm.

(17) 5-sided (18) 8-sided (19) 10-sided (20) 12-sided.

(21) Draw a circle radius 4 cm and mark off the circumference into 8 equal parts. Join four of the points you have marked to form a square. Join the other 4 points to make another square. Join the corners of the squares to form an octagon.
(22) Repeat question 21, drawing two pentagons to form a decagon.
(23) Repeat question 21, drawing two equilateral triangles to form a hexagon.
(24) Repeat question 21, drawing two hexagons to form a dodecagon.
(25) Draw two concentric circles, with radii 2.5 cm and 5 cm. Divide the circumference of the larger circle into 18 equal parts and draw an 18-sided figure in this circle. Use your construction lines to draw a 9-sided figure in the small circle.

Exercise 26b

(1) (a) What is the name given to a 5-sided figure?
(b) What is the size of the exterior angle of a regular 20-sided figure?

(c) What is the size of each interior angle of a regular 20-sided figure?

(d) Draw a regular octagon in a circle radius 5 cm and measure the side of your octagon, correct to 0.1 cm.

(2) (a) What is the name given to a 10-sided figure?

(b) What is the sum of the interior angles of a regular pentagon?

(c) What is the size of the exterior angle of a regular 36-sided figure?

(d) Draw a regular hexagon in a circle radius 4 cm.

(3) (a) How many sides has a regular octagon?

(b) What is the sum of the interior angles of an n-sided figure?

(c) Calculate the exterior angle of a regular 9-sided figure.

(d) Draw a regular 9-sided figure and measure the interior angle of your figure.

(4) (a) What is the name given to a regular quadrilateral?

(b) What is the size of the exterior angle of an equilateral triangle?

(c) Which regular polygon is made up of equilateral triangles?

(d) Draw a regular pentagon in a circle radius 4 cm and measure the side of your pentagon correct to 0.1 cm.

Section E
27 Indices

Negative Indices

To divide x^5 by x^3, we can simply subtract the indices and say that $x^5 \div x^3 = x^2$.

Can we still use the same method for $x^5 \div x^8$? There are two ways in which this can be simplified.

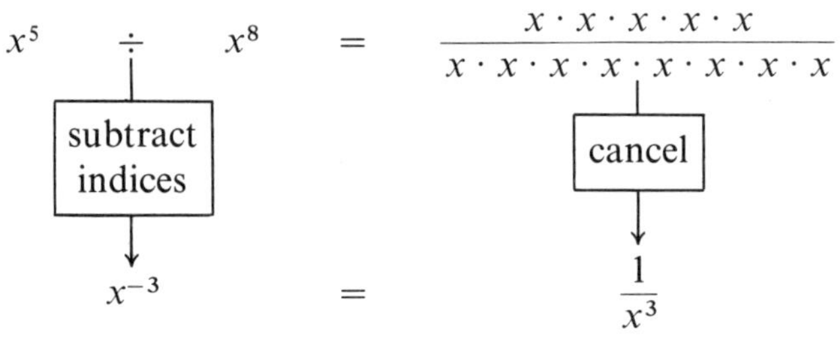

$$x^5 \div x^8 = \frac{x \cdot x \cdot x \cdot x \cdot x}{x \cdot x \cdot x \cdot x \cdot x \cdot x \cdot x \cdot x}$$

subtract indices ↓   cancel ↓

$$x^{-3} = \frac{1}{x^3}$$

This shows us that we can subtract the indices if we take x^{-3} to mean $\frac{1}{x^3}$. In the same way $x^{-2} = \frac{1}{x^2}$, $x^{-5} = \frac{1}{x^5}$ and so on.

Zero Index

We now know that $x^m \div x^n$ can be simplified if $m > n$ or if $m < n$. What happens if $m = n$? Suppose $m = 3$.

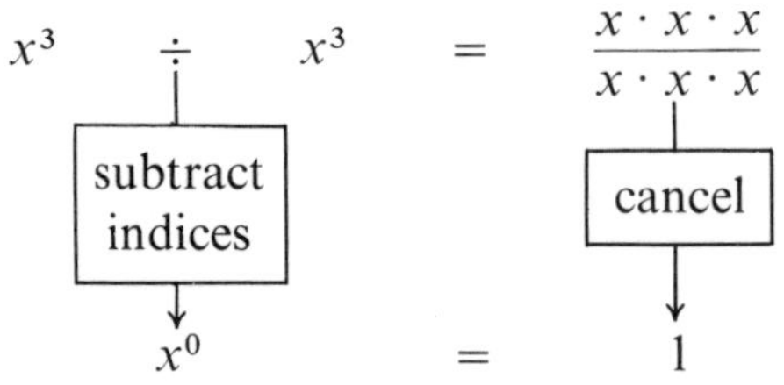

$$x^3 \div x^3 = \frac{x \cdot x \cdot x}{x \cdot x \cdot x}$$

subtract indices ↓   cancel ↓

$$x^0 = 1$$

This shows us that x^0 is to be taken as 1, for all values of x.

Example: Simplify
(i) $x^3 \times x^{-4}$ (ii) $y^2 \div y^5$.

1. Multiply by adding indices.

(i) $x^3 \times x^{-4} = x^{-1}$.

2. Divide by subtracting indices.

(ii) $y^2 \div y^5 = y^{-3}$.

Fraction Indices

We now know that indices can have any whole number value, but what about fractions? Is there any meaning for $x^{\frac{1}{2}}$? If we can use $x^{\frac{1}{2}}$ in the way that we use other indices, then $x^{\frac{1}{2}} \times x^{\frac{1}{2}} = x^1 = x$. We also know that $\sqrt{x} \times \sqrt{x} = x$.

$$x^{\frac{1}{2}} \times x^{\frac{1}{2}} = x^1$$
$$\downarrow$$
$$\sqrt{x} \times \sqrt{x} = x.$$

This shows us that if $x^{\frac{1}{2}}$ is taken to mean $\sqrt{x}$, then it can be used like other indices.

Example: Find the value of
(i) x^{-3} if $x = 6$
(ii) $x^{\frac{1}{2}}$ if $x = 144$
(iii) x^0 if $x = 17$.

1. Change to a positive index.

(i) $x^{-3} = \dfrac{1}{x^3}$

2. Put 6 instead of x.

$= \dfrac{1}{6^3}$

3. Find the value of 6^3.

$= \dfrac{1}{216}$.

1. $x^{\frac{1}{2}}$ means $\sqrt{x}$.
2. $12^2 = 144$, so $\sqrt{144} = 12$.

(ii) $x^{\frac{1}{2}} = \sqrt{x}$

$= 12$.

(iii) $x^0 = 1$ for all values of x.

When multiplying or dividing with negative, zero, or fraction indices, they should be used in just the same way as indices which are positive whole numbers.

Example: Simplify $3x^2 \times 4x^{\frac{1}{2}}$.

1. Multiply coefficients.

$3x^2 \times 4x^{\frac{1}{2}} = 12■$

2. Multiply the x parts by adding indices. ($2 + \frac{1}{2} = 2\frac{1}{2} = \frac{5}{2}$.)

$= 12x^{\frac{5}{2}}$.

In this example, note that $x^{2\frac{1}{2}}$ means $x^2 \times x^{\frac{1}{2}}$ or $x^2\sqrt{x}$, and that $x^{2\frac{1}{2}}$ is usually written as $x^{\frac{5}{2}}$.

Square Roots

The fact that $\sqrt{x} = x^{\frac{1}{2}}$ shows us that square roots can be found by halving the indices. Thus $\sqrt{(x^6)} = x^3$, $\sqrt{(z^{10})} = z^5$ and so on.

Example: Write down the square root of $25x^8y^6$.

1. Write the example.

$\sqrt{(25x^8y^6)} = ■$

2. Write down $\sqrt{25}$.

$\sqrt{(25x^8y^6)} = 5■$

3. Find $\sqrt{(x^8)}$ by halving the index. ($8 \div 2 = 4$.)

$\sqrt{(25x^8y^6)} = 5x^4■$

4. Find $\sqrt{(y^6)}$ by halving the index. ($6 \div 2 = 3$.)

$\sqrt{(25x^8y^6)} = 5x^4y^3$.

Standard Notation

The earth is roughly 150 000 000 km from the sun. This and other large numbers can be written without all the zeros by using indices.

$150\,000\,000 = 1.5 \times 10^8$.

In the same way, indices can be used to write very small numbers.

$$0.000\,000\,005 = 5.0 \div 10^9$$
$$= 5.0 \times 10^{-9}.$$

Exercise 27a

Simplify

(1) $x^3 \div x^7$
(2) $x^4 \div x^9$
(3) $x^2 \div x^2$
(4) $x^6 \div x^{10}$
(5) $y^4 \div y^8$
(6) $p^8 \div p^{10}$
(7) $z^7 \div z^7$
(8) $x^9 \div x^{12}$
(9) $q^3 \div q^8$
(10) $s^2 \div s^7$
(11) $x^3 \times x^{-1}$
(12) $y^2 \times y^{-4}$
(13) $s^6 \times s^{-2}$
(14) $x^7 \cdot x^{-10}$
(15) $z^5 \times z^{-5}$
(16) $x^2 \div x^{-3}$
(17) $x^3 \div x^{-4}$
(18) $x^{\frac{1}{2}} \cdot x$
(19) $x^2 \div x^{\frac{1}{2}}$
(20) $x^{\frac{1}{2}} \div x^{\frac{1}{2}}$

Write the following using negative indices

(21) $\frac{1}{x^2}$ (22) $\frac{1}{y^4}$ (23) $\frac{1}{p^9}$ (24) $\frac{1}{z^8}$ (25) $\frac{1}{a^3}$

Write the following using positive indices

(26) x^{-6} (27) y^{-3} (28) z^{-1} (29) p^{-7} (30) z^{-4}

Find the value of each of the following:

(31) x^{-1}, if $x = 11$
(32) x^{-5}, if $x = 2$
(33) z^{-2}, if $z = 12$
(34) p^0, if $p = 8$
(35) $x^{\frac{1}{2}}$, if $x = 9$
(36) $z^{\frac{1}{2}}$, if $z = 4$
(37) x^{-2}, if $x = 9$
(38) x^0, if $x = 3$
(39) $a^{\frac{1}{2}}$, if $a = 25$
(40) x^{-2}, if $x = 1$.

Write down the square root of each of the following:

(41) $4x^4$
(42) $9x^6$
(43) $64x^4$
(44) $81x^8$
(45) $25x^6$
(46) 9×10^6
(47) 4×10^4
(48) 25×10^8
(49) 16×10^2
(50) 9×10^{12}

Write each of the following using the standard notation:

(51) 2 700 000
(52) 3 100
(53) 42 000 000
(54) 570
(55) 9 400 000
(56) 0.34
(57) 0.0087
(58) 0.000 003 2
(59) 0.000 46
(60) 0.000 000 091.

Exercise 27b

(1) (a) If $a = 2$, $b = 3$ and $c = 4$, calculate the value of a^3b^2c.
(b) Write down the square root of $25x^6y^{10}$.
(c) Simplify $12xy^3 \div 4x^4y^2$.

(d) If the length of a paddock is 33 m and its width is 9.7 m, calculate the area in square metres. Give your answer in standard form correct to 2 significant figures.

(2) If $A = 3x^2 + 1$ and $B = (3x + 1)^2$, find the value of

(a) $A \times B$ if $x = 2$

(b) $B \div A$ if $x = 1$

(c) B^{-1} if $x = 3$ (Give your answer in standard form.)

(d) A^0 if $x = 5$.

(3) If $P = 4x + 2$ and $Q = 5x + 3$, find

(a) PQ in descending powers of x

(b) the value of x if $P = Q$

(c) the value of Q^2 if $x = 5$

(d) the value of P^{-3} if $x = \frac{1}{2}$.

(4) If $p = 25$, $q = 3$ and $r = -1$, calculate the value of

(a) qr^2

(b) $p^{\frac{1}{2}}$.

Simplify the following:

(c) $3x^2y^4 \times 5x^{-4}y^3$

(d) $4x^3 \div 5x^2y^2$.

(5) An aeroplane travels east for 2 hours at 400 km/h and north for 3 hours at 500 km/h.

(a) Calculate the total distance travelled and give your answer in standard form. Make a map of the journey using a scale of 1 cm to 100 km and from your map, find (b) the distance between the starting point and the finishing point of the journey. (Give your answer in standard form.)

(c) Find the course which the pilot would have to set if he were to fly directly back to his starting point.

(d) Calculate how long this direct journey back would take at an average speed of 400 km/h.

28 Square Roots

Square Root of a Number

If $y^2 = x$, we say that y is the *square root* of x, and we write $y = \sqrt{x}$. Some whole numbers are perfect squares, so that the square root is also a whole number.

$$\sqrt{1} = 1 \qquad \sqrt{9} = 3$$
$$\sqrt{4} = 2 \qquad \sqrt{16} = 4.$$

Most square roots are not whole numbers, and are found from square root tables. These tables are not perfectly accurate. They give square roots correct to 3 or 4 significant figures. In fact, a number such as $\sqrt{2}$ cannot be written accurately. Like π, it is a never-ending decimal.

Square Root Tables

We know that

$$2^2 = 4$$
$$20^2 = 400 = 4.0 \times 10^2$$
$$200^2 = 40\,000 = 4.0 \times 10^4.$$

This means that we can find $\sqrt{400}$ by looking up $\sqrt{4}$ in the tables

$$\sqrt{4} = 2$$
$$\sqrt{400} = 2 \times 10$$
$$\sqrt{40\,000} = 2 \times 10^2.$$

To find $\sqrt{40}$ is more difficult, because

$$\sqrt{40} \simeq 6.3 \qquad (\sqrt{40} \text{ is approximately } 6.3)$$
$$\sqrt{4\,000} \simeq 6.3 \times 10 \qquad (\sqrt{4\,000} \text{ is approximately } 6.3 \times 10)$$
$$\sqrt{400\,000} \simeq 6.3 \times 10^2 \qquad (\sqrt{400\,000} \text{ is approximately } 6.3 \times 10^2).$$

This shows that we can find $\sqrt{4000}$ by looking up $\sqrt{40}$ in the tables but *not* by looking up $\sqrt{4}$. Because of this, square root tables must have the square roots of all numbers from 1 to 100. The square root tables usually take up two double pages in a book of tables. The first part gives square roots of numbers from 1 to 10 and the second part gives square roots of numbers from 10 to 100.

Four-Figure Tables

The programme starting on page 131 shows how to use four-figure square root tables.

SQUARE ROOTS OF NUMBERS 1–10

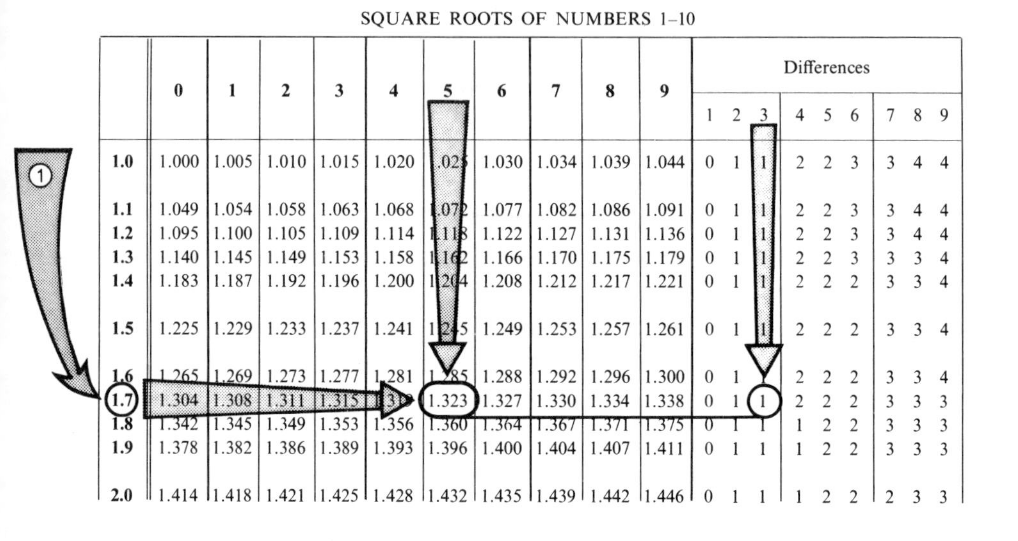

	0	1	2	3	4	5	6	7	8	9	Differences								
											1	2	3	4	5	6	7	8	9
1.0	1.000	1.005	1.010	1.015	1.020	1.025	1.030	1.034	1.039	1.044	0	1	1	2	2	3	3	4	4
1.1	1.049	1.054	1.058	1.063	1.068	1.072	1.077	1.082	1.086	1.091	0	1	1	2	2	3	3	4	4
1.2	1.095	1.100	1.105	1.109	1.114	1.118	1.122	1.127	1.131	1.136	0	1	1	2	2	3	3	4	4
1.3	1.140	1.145	1.149	1.153	1.158	1.162	1.166	1.170	1.175	1.179	0	1	1	2	2	3	3	3	4
1.4	1.183	1.187	1.192	1.196	1.200	1.204	1.208	1.212	1.217	1.221	0	1	1	2	2	2	3	3	4
1.5	1.225	1.229	1.233	1.237	1.241	1.245	1.249	1.253	1.257	1.261	0	1	1	2	2	2	3	3	4
1.6	1.265	1.269	1.273	1.277	1.281	1.285	1.288	1.292	1.296	1.300	0	1	1	2	2	2	3	3	4
1.7	1.304	1.308	1.311	1.315	[illegible]	1.323	1.327	1.330	1.334	1.338	0	1	1	2	2	2	3	3	3
1.8	1.342	1.345	1.349	1.353	1.356	1.360	1.364	1.367	1.371	1.375	0	1	1	1	2	2	3	3	3
1.9	1.378	1.382	1.386	1.389	1.393	1.396	1.400	1.404	1.407	1.411	0	1	1	1	2	2	3	3	3
2.0	1.414	1.418	1.421	1.425	1.428	1.432	1.435	1.439	1.442	1.446	0	1	1	1	2	2	2	3	3

SQUARE ROOTS OF NUMBERS 10–100

	0	1	2	3	4	5	6	7	8	9	Differences								
											1	2	3	4	5	6	7	8	9
10	3.162	3.178	3.194	3.209	3.225	3.240	3.256	3.271	3.286	3.302	2	3	5	6	8	9	11	12	14
11	3.317	3.332	3.347	3.362	3.376	3.391	3.406	3.421	3.435	3.450	1	3	4	6	7	9	10	12	13
12	3.464	3.479	3.493	3.507	3.521	3.536	3.550	3.564	3.578	3.592	1	3	4	6	7	8	10	11	13
13	3.606	3.619	3.633	3.647	3.661	3.674	3.688	3.701	3.715	3.728	1	3	4	5	7	8	10	11	12
14	3.742	3.755	3.768	3.782	3.795	3.808	3.821	3.834	3.847	3.860	1	3	4	5	7	8	9	11	12
15	3.873	3.886	3.899	3.912	3.924	3.937	3.950	3.962	3.975	3.987	1	3	4	5	6	8	9	10	11
16	4.000	4.012	4.025	4.037	4.050	4.062	4.074	4.087	4.099	4.111	1	2	4	5	6	7	9	10	11
17	4.123	4.135	4.147	4.159	4.171	4.183	4.195	4.207	4.219	4.231	1	2	4	5	6	7	8	10	11
18	4.243	4.254	4.266	4.278	4.290	4.301	4.313	4.324	4.336	4.347	1	2	3	5	6	7	8	9	10
19	4.359	4.370	4.382	4.393	4.405	4.416	4.427	4.438	4.450	4.461	1	2	3	5	6	7	8	9	10
20	4.472	4.483	4.494	4.506	4.517	4.528	4.539	4.550	4.561	4.572	1	2	3	4	6	7	8	9	10

Answers ↓	
	Let us find the square root of 1.753. Look at the table opposite. Look at the part headed 'Square roots of numbers 1–10'. Follow the number 1 arrow down the first column to 1.7 and move your finger across the page until you get to the 5 column. The number has a ring round it. What number is it?
1.323	$1.323 = \sqrt{1.75}$. We want $\sqrt{1.753}$ so we must add the number in the 3 difference column. Keep your finger on 1.323 and move another finger across to the 3 difference column. What number do you find?
1	Add the 1.323 and the 1 as shown below $\sqrt{1.75}$ → 1.323 from 3 difference column → 1 1.323 1 1.324 What is the value of $\sqrt{1.753}$?
1.324 (correct to 4 significant figures)	Look at the table opposite. Look at the part headed 'Square roots of numbers 10–100'. Can you find $\sqrt{17.53}$? Run your finger down the first column until you get to 17. Move your finger across to the 5 column. What number do you find there?
4.183	Keep your finger on 4.183 and move another finger across to the 3 difference column. What number do you find there?
4	Add 4.183 and 4 as before. What is the value of $\sqrt{17.53}$?
4.187	Now open your book of tables at the page headed 'Square roots of numbers 1–10'. What is the value of $\sqrt{3.6}$?
1.897	What is the value of $\sqrt{2.54}$?
1.594	What is the value of $\sqrt{4.977}$?
2.231	Now open your tables to the page headed 'Square roots of numbers 10–100'. What is the value of $\sqrt{37}$?
6.083	What is the value of $\sqrt{44.6}$?

6.678	What is the value of $\sqrt{51.32}$?
7.163	From now on, use your own square root tables to find all square roots. You will not be told which page to look on. You must decide this for yourself. What is the value of $\sqrt{7.392}$?
2.718	What is the value of $\sqrt{5.663}$?
2.380	What is the value of $\sqrt{67.45}$?
8.213	Now do Exercise 28a, questions 1–30, before going on to the next piece of work.

Which Table?

There are two square root tables. One gives square roots of numbers from 1–10, the other gives square roots of numbers from 10–100. Before you can find the square root of a number you must decide which table to look in.

We know from algebra that to find $\sqrt{(4x^6)}$ we must find $\sqrt{4}$ which is 2. We next find $\sqrt{(x^6)}$ by halving the index $(6 \div 2 = 3)$. Then,

$$\sqrt{(4x^6)} = 2x^3.$$

If $x = 10$, we have $\sqrt{(4 \times 10^6)} = 2 \times 10^3$.

This helps us to find the square root of any number which is greater than 100.

The method can also be used for negative indices, which helps us to find the square root of any number less than 1.

$$\sqrt{(25x^{-6})} = 5x^{-3}.$$

If $x = 10$, $\sqrt{(25 \times 10^{-6})} = 5 \times 10^{-3}$

so that $\sqrt{0.000\,025} = 0.005$.

Notice that the square root is larger than the number we started with. This always happens for numbers less than 1. For example, $\frac{1}{2} \times \frac{1}{2} = \frac{1}{4}$. Therefore, $\sqrt{\frac{1}{4}} = \frac{1}{2}$. We also know that $\frac{1}{2} = 0.5$ and that $(0.5)^2 = 0.25$, so that $\sqrt{0.25} = 0.5$.

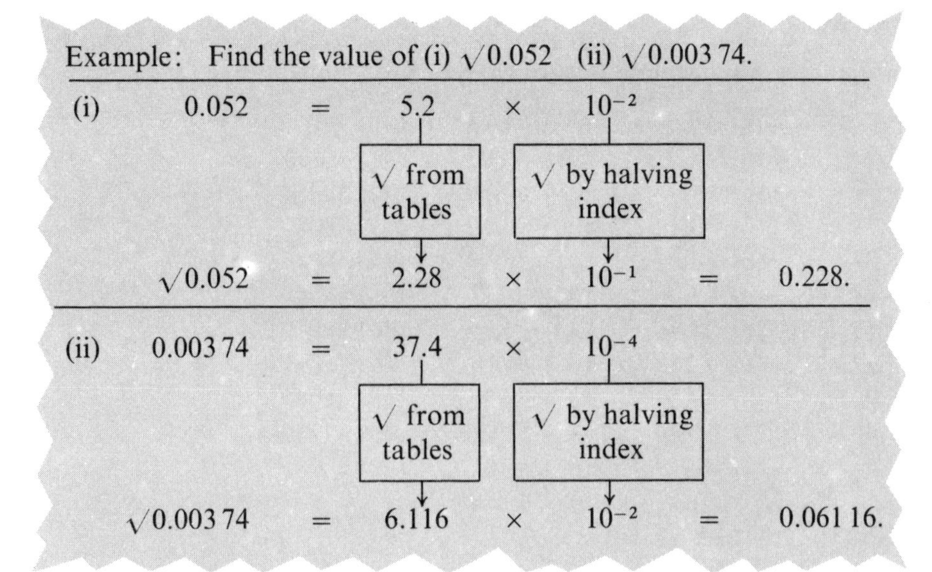
Example: Find the value of (i) $\sqrt{0.052}$ (ii) $\sqrt{0.00374}$.

(i) $0.052 = 5.2 \times 10^{-2}$

$\sqrt{0.052} = 2.28 \times 10^{-1} = 0.228.$

(ii) $0.003\,74 = 37.4 \times 10^{-4}$

$\sqrt{0.003\,74} = 6.116 \times 10^{-2} = 0.061\,16.$

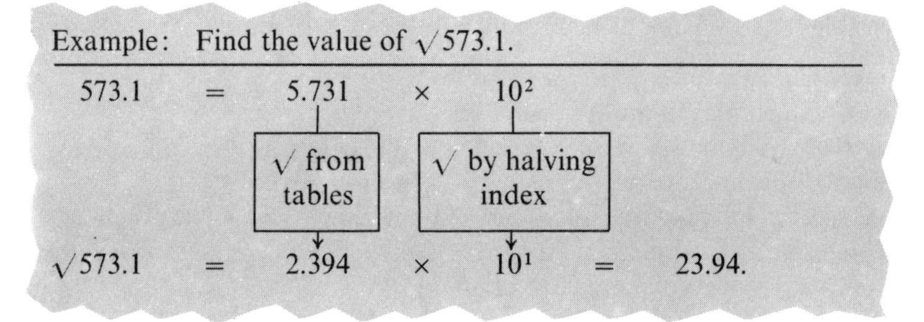
Example: Find the value of $\sqrt{573.1}$.

$573.1 = 5.731 \times 10^{2}$

$\sqrt{573.1} = 2.394 \times 10^{1} = 23.94.$

In order to use this method, the number must first be put in the form $x \times 10^n$, where x is a number between 1 and 100, and n is an even number.

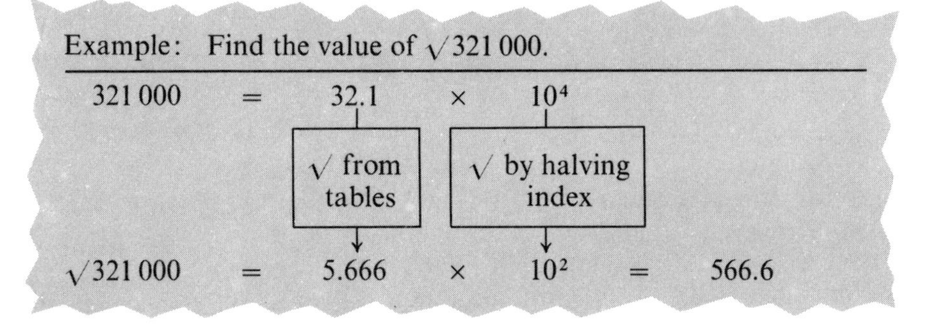
Example: Find the value of $\sqrt{321\,000}$.

$321\,000 = 32.1 \times 10^{4}$

$\sqrt{321\,000} = 5.666 \times 10^{2} = 566.6$

Exercise 28a

Use tables to find the square root of each of the following:

(1) 5.7	(14) 35.9	(27) 81.48	(40) 961.3
(2) 3.2	(15) 48.2	(28) 3.793	(41) 4 040
(3) 9.5	(16) 64.7	(29) 67.29	(42) 0.034 2
(4) 4.71	(17) 10.1	(30) 96.62	(43) 0.082 13
(5) 8.84	(18) 53.01	(31) 562.4	(44) 0.084 2
(6) 6.72	(19) 42.78	(32) 9 327	(45) 0.149
(7) 7.36	(20) 94.36	(33) 342.9	(46) 0.010 7
(8) 5.672	(21) 6.28	(34) 452.1	(47) 0.041 4
(9) 8.921	(22) 59.7	(35) 3 026	(48) 0.061 8
(10) 9.451	(23) 41.84	(36) 521.3	(49) 0.239
(11) 47	(24) 4.61	(37) 8 161	(50) 0.081 7.
(12) 89	(25) 5.63	(38) 3 274	
(13) 32	(26) 32.32	(39) 842	

Exercise 28b

(1) (a) Write down the square root of $49a^2b^4$.

(b) Use your tables to find the value of $\sqrt{}0.053$.

(c) If $P = \sqrt{}(2ab)$, where $a = 9$ and $b = 11$, calculate the value of P correct to 1 decimal place.

(d) A square has an area of 472 m². Find the length of the side correct to 3 significant figures.

(2) (a) Write down the square root of 64×10^6.

(b) Use your tables to find the value of $\sqrt{}37\,200$.

(c) Solve the equation $x^2 = 2\,340$.

(d) In the figure, ABCD is a square. E, F, G and H are the mid-

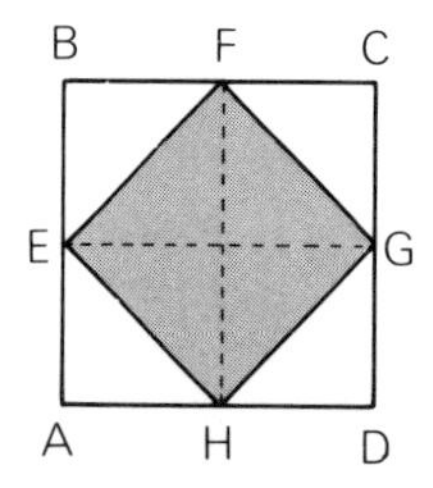

points of AB, BC, CD and DA. If the area of ABCD = 423 cm² calculate the length of EF. (First find the area of EFGH.)

(3) (a) Write down the square root of $81x^6y^8$.

(b) Write down the square root of 36×10^{-6}.

(c) If $c = \sqrt{a^2 + b^2}$, calculate the value of c when $a = 7$ and $b = 12$. Give your answer correct to 3 significant figures.

(d) A rectangle has length 23 cm and width 14 cm. A square ABCD has the same area as the rectangle. Calculate the length of AB to 1 decimal place.

(4) (a) Use your tables to find the value of $\sqrt{0.324}$.

(b) Solve the equation $x^2 = 130$ correct to 3 significant figures.

(c) A square prism has a cross-sectional area A and length l. Calculate A if $l = 10$ and the volume is 123 cm³ ($V = Al$).

(d) If the prism in (c) has cross-section ABCD, calculate the length of AB correct to 2 decimal places.

(5) (a) Write down the square root of $9x^{-4}$.

(b) Find the set of prime factors of 1 764.

(c) Use your answer to (b) to find the square root of 1 764.

(d) In $\triangle$ABC, $a = 5$ cm, $b = 6$ cm and $\angle$C $= 90°$. Use the Pythagoras rule ($a^2 + b^2 = c^2$) to find c^2 and calculate c correct to 2 significant figures.

29 Logarithms

Bar Numbers

When we divide two numbers using logs, we sometimes end up with a logarithm with a bar number. This simply means that it is the logarithm of a number which is less than 1.

Example: Calculate $43.72 \div 93.5$.

No	Log
43.72	1.6407
93.50	1.9708
0.4677	$\bar{1}.6699$

$43.72 \div 93.5 = 0.468$
(correct to 3 significant figures).

Adding Bar Numbers

If we multiply or divide by numbers less than 1, we start off with bar numbers. It is important to remember that a bar number is negative. This means that the whole number part of the log is negative, but the decimal part is positive.

Example: Add $\bar{3}.2 + \bar{1}.4$.

1. Add the decimal parts $(2 + 4 = 6)$.

2. Add the whole number parts $(\bar{3} + \bar{1} = \bar{4})$.

$$\begin{array}{l} \bar{3}.2 \\ \underline{\bar{1}.4} + \\ \underline{\bar{4}.6} \end{array}$$

In the last example, the numbers could be added easily, because the positive and negative parts were quite separate. This is not always the case.

Example: Add $\bar{4}.5 + \bar{2}.7$.

$\bar{4}.5\ +$ $\bar{2}.7$ ■.2 1	$\bar{4}.5\ +$ $\bar{2}.7$ $\bar{5}.2$ 1
1. Add the decimal parts $(5 + 7 = 12)$. Write the 2 and transfer the 1.	2. Add the whole number parts $(\bar{4} + \bar{2} = \bar{6}, \bar{6} + 1 = \bar{5})$.

If some of the whole number parts are positive and some negative, add them separately.

Example: Add $\bar{4}.5 + 2.6 + \bar{3}.7$.

$\bar{4}.5$ 2.6 $\bar{3}.7$ ■.8 1	$\bar{4}.5$ 2.6 $\bar{3}.7$ $\bar{4}.8$ 1
1. Add the decimal parts $(5 + 6 + 7 = 18)$. Write the 8 and transfer the 1.	2. Add the positive whole numbers $(2 + 1 = 3)$. Add the negative whole numbers $(\bar{4} + \bar{3} = \bar{7})$. Add positives and negatives $(3 + \bar{7} = \bar{4})$.

Subtracting Bar Numbers

Before we can understand how to subtract bar numbers, we must first look at how to subtract positive numbers. For example, $5.3 - 1.7 = 3.6$.

$$\begin{array}{r} 5.3 \\ -\ 1.7 \\ \hline \blacksquare \end{array} \longrightarrow \begin{array}{r} 4.{}^{1}3 \\ -\ 1.\ 7 \\ \hline \blacksquare \end{array} \longrightarrow \begin{array}{r} 4.{}^{1}3 \\ -\ 1.\ 7 \\ \hline 3.\ 6 \end{array}$$

To do this we say that $5.3 = 4.0 + 1.3$. We take 1 from the 5 ($5 - 1 = 4$) and transfer it to the decimal column ($10 + 3 = 13$). Negatives can be subtracted in the same way.

$$\begin{array}{r} \bar{5}.3 \\ -\ \bar{1}.7 \\ \hline \blacksquare \end{array} \longrightarrow \begin{array}{r} \bar{6}.{}^{1}3 \\ -\ \bar{1}.\ 7 \\ \hline \blacksquare \end{array} \longrightarrow \begin{array}{r} \bar{6}.{}^{1}3 \\ -\ \bar{1}.\ 7 \\ \hline \bar{5}.\ 6 \end{array}$$

In this case, we take 1 from the $\bar{5}$ ($\bar{5} - 1 = \bar{6}$) and transfer it to the decimal as before ($10 + 3 = 13$).

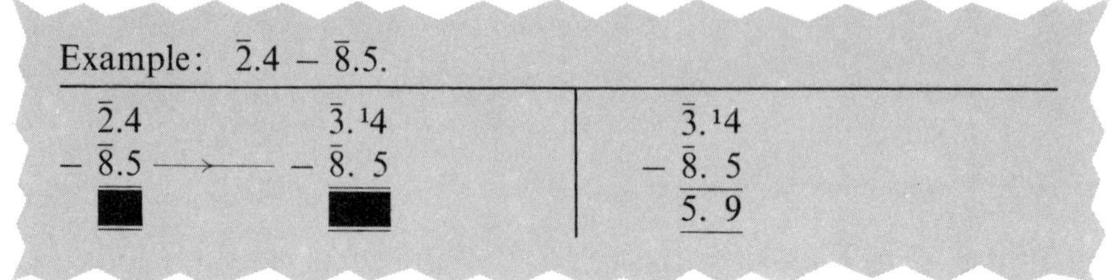

1. $4 < 5$, so transfer 1 from $\bar{2}$ ($\bar{2} - 1 = \bar{3}$) to the decimal part ($10 + 4 = 14$).

2. Subtract the decimal parts ($14 - 5 = 9$).
Subtract the whole numbers ($\bar{3} - \bar{8} = 5$).

Using Bar Numbers

The following example shows how bar numbers are used in calculations with logarithms.

Example: Calculate $\dfrac{42.8 \times 0.0325}{0.273 \times 0.009}$ correct to 3 significant figures.

No	Log
42.8	1.6314
0.0325	$\bar{2}.5119$
	0.1433
0.273	$\bar{1}.4362$
0.009	$\bar{3}.9542$
	$\bar{3}.3904$
566.1	2.7529

$$\frac{42.8 \times 0.0325}{0.273 \times 0.009} = 566$$

(correct to 3 significant figures).

Exercise 29a

Calculate the following:

(1) $\bar{1}.2 + \bar{3}.7$
(2) $\bar{3}.2 + \bar{2}.4$
(3) $\bar{2}.3 + 3.4$
(4) $\bar{2}.4 + 4.3$
(5) $3.6 + \bar{1}.2$
(6) $\bar{3}.2 + \bar{1}.3$
(7) $\bar{2}.1 + 1.7$
(8) $\bar{4}.1 + 3.5$
(9) $1.1 + \bar{2}.7$
(10) $\bar{1}.3 + 2.8$
(11) $\bar{3}.7 + 4.5$
(12) $3.9 + \bar{2}.5$
(13) $\bar{4}.8 + 2.8$
(14) $\bar{3}.6 + \bar{1}.5$
(15) $\bar{2}.4 + 0.7$
(16) $\bar{3}.2 + \bar{1}.9$
(17) $\bar{1}.6 + \bar{2}.7$
(18) $\bar{3}.1 + \bar{1}.9$
(19) $\bar{1}.4 + \bar{2}.8$
(20) $\bar{1}.7 + 0.9$
(21) $\bar{3}.6 - 2.4$
(22) $\bar{1}.5 - 0.2$
(23) $\bar{4}.3 - 2.1$
(24) $\bar{2}.7 - 3.5$
(25) $2.7 - \bar{3}.5$
(26) $3.2 - \bar{1}.1$
(27) $4.6 - \bar{2}.3$
(28) $4.1 - \bar{2}.7$
(29) $\bar{2}.3 - \bar{3}.1$
(30) $\bar{1}.6 - \bar{2}.3$
(31) $\bar{3}.4 - \bar{4}.2$
(32) $\bar{3}.3 - \bar{1}.2$
(33) $\bar{4}.7 - \bar{2}.3$
(34) $\bar{5}.4 - \bar{1}.1$
(35) $2.1 - \bar{1}.3$
(36) $3.2 - \bar{2}.5$
(37) $5.3 - \bar{1}.6$
(38) $1.2 - \bar{2}.7$
(39) $\bar{2}.3 - 1.6$
(40) $\bar{1}.7 - 1.8$
(41) $\bar{2}.1 - 1.9$
(42) $\bar{4}.5 - 2.7$
(43) $\bar{3}.4 - \bar{4}.7$
(44) $\bar{1}.6 - \bar{3}.8$
(45) $\bar{2}.5 - \bar{2}.8$
(46) $\bar{3}.4 - \bar{2}.9$
(47) $\bar{3}.4 - \bar{1}.7$
(48) $\bar{5}.2 - \bar{2}.9$
(49) $\bar{4}.6 - \bar{1}.8$
(50) $\bar{3}.4 - \bar{2}.7$
(51) 3.65×0.712
(52) 0.02×5.71
(53) 3.92×0.45
(54) 0.34×0.25
(55) 0.0035×4.72
(56) $57.2 \div 0.34$
(57) $2.96 \div 0.97$
(58) $0.02 \div 4.3$
(59) $0.356 \div 0.015$
(60) $0.141 \div 11.3$.

Exercise 29b

Give all answers correct to 3 significant figures.

(1) Calculate
(a) 32.5×0.431
(b) $4.32 \div 0.254$
(c) $24.72 \times 3.01 \times 0.016$
(d) $\dfrac{0.26 \times 0.481}{0.094}$.

(2) Calculate
(a) 0.326×0.15
(b) $38.6 \div 0.035$
(c) $\dfrac{3.148 \times 0.76}{1.09 \times 0.77}$
(d) the volume of a match stick which is 2.3 cm long, if its cross-section is a square of side 0.15 cm.

(3) Calculate
(a) $1.033 \div 0.34$

(b) $(0.351)^2$

(c) the area of a rectangular piece of card which is 0.47 m long and 0.253 m wide,

(d) the distance travelled in 0.72 hours at an average speed of 43.2 km/h.

(4) Calculate

(a) 0.73×0.451

(b) $0.634 \div 0.072$

(c) the volume of a prism with cross-sectional area 0.932 cm² and length 0.73 cm. ($V = Al$.)

(d) the area of a circle of radius 0.155 m. ($A = \pi r^2$, take $\pi = 3.142$.)

(5) (a) Calculate $\dfrac{3.25 \times 0.79}{0.0414}$.

(b) Find the cost in £ to the nearest penny of 11.35 dollars, if 1 dollar costs £0.42.

(c) If $P = \dfrac{x}{y}$, find the value of P when $x = 0.376$ and $y = 0.054$.

(d) Calculate the circumference of a circle with radius 0.729 m. ($C = 2\pi r$, take $\pi = 3.142$.)

30 Solids

Faces, Vertices and Edges

A solid figure with no curved surface is called a *polyhedron*. (Say *polyhedra* for more than one.) Here are some polyhedra.

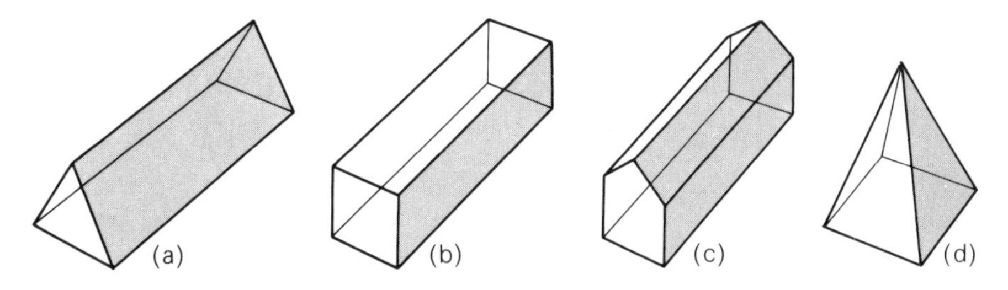

Each surface of a polyhedron is a *plane face*. In other words, it is not curved. Two faces of a polyhedron meet in a line called an *edge*. Three or more edges meet at a point called a *vertex*. (Vertex is another word for corner. Say *vertices* for more than one.)

Count the number of faces, edges and vertices for the solids shown above. Copy and complete the table below.

solid	No. of faces	No. of vertices	No. of edges	$F + V - E$
(a)				
(b)				
(c)				
(d)				

If F is the number of faces, V the number of vertices and E the number of edges, you should find that

$$F + V - E = 2.$$

A solid with a constant cross-section is called a *prism*. In the figure above, (a), (b) and (c) are all prisms, (d) is a square *pyramid*.

Regular Solids

A polyhedron with all faces the same shape and size and each one making the same angle with the others is called a *regular solid*. There are five regular solids, and these are shown below.

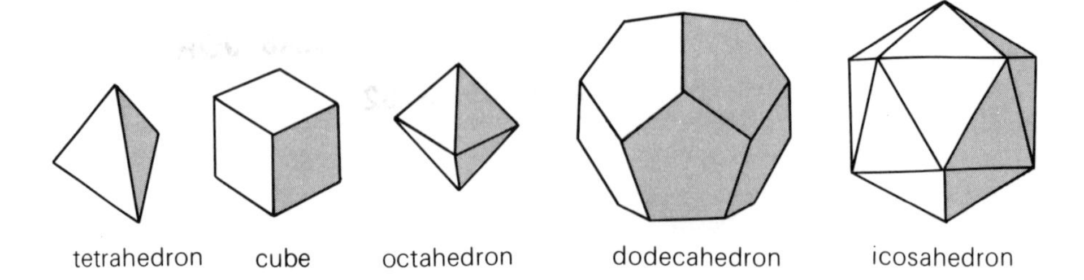

The table shows the number of faces for each regular solid, and the shape of the faces in each case.

solid	number of faces	shape of faces
tetrahedron	4	equilateral triangle
cube	6	square
octahedron	8	equilateral triangle
dodecahedron	12	regular pentagon
icosahedron	20	equilateral triangle

If a hollow shape is opened out, the result is the *net* of the solid. The net of each of the regular solids is shown below. If these are made from thin card, tabs will be needed and these are shown by the shaded parts.

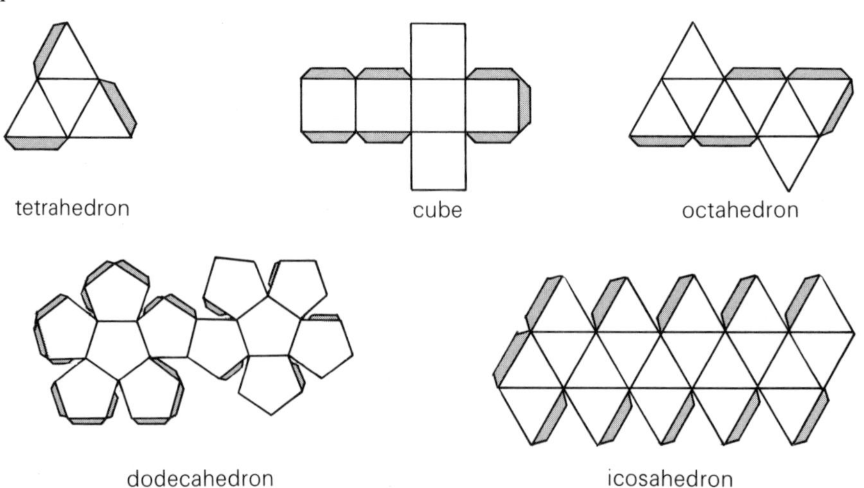

Curved Solids

The most important curved solids are the cylinder, cone, and sphere.

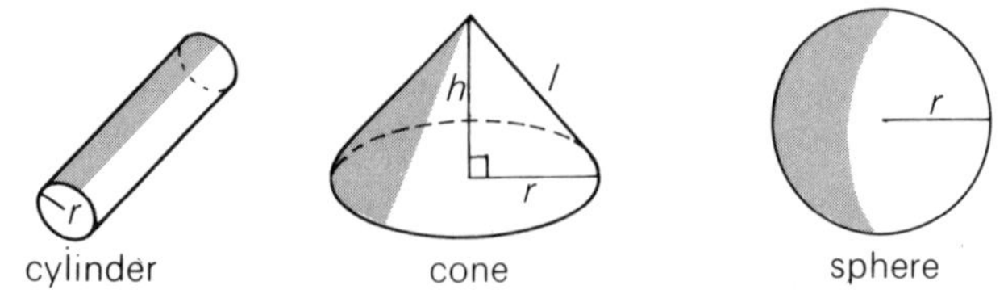

A cylinder is a circular prism. When we talk of the radius of a cylinder we mean the radius of its cross-section. When we talk of the base radius of a cone we mean the radius of its base. The distance marked l in the figure is called the *slant height* of the cone. Half a sphere is called a *hemisphere*.

If the top of a cone is cut off parallel to the base, the part which is left is called a *frustum*.

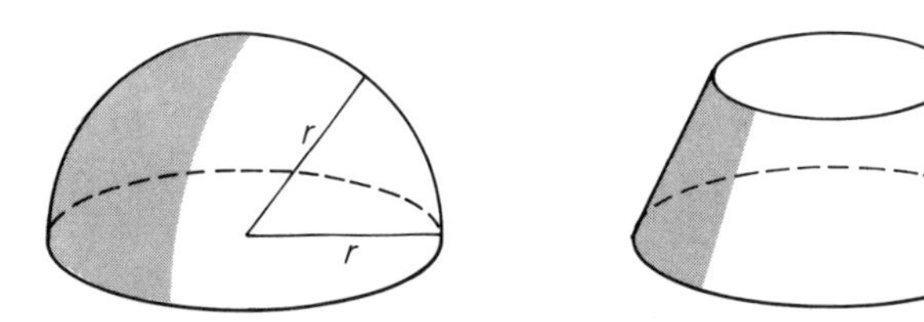

A cross-section of a sphere is a circle and a horizontal cross-section of a cone is also a circle.

The nets of a cylinder and cone are shown below.

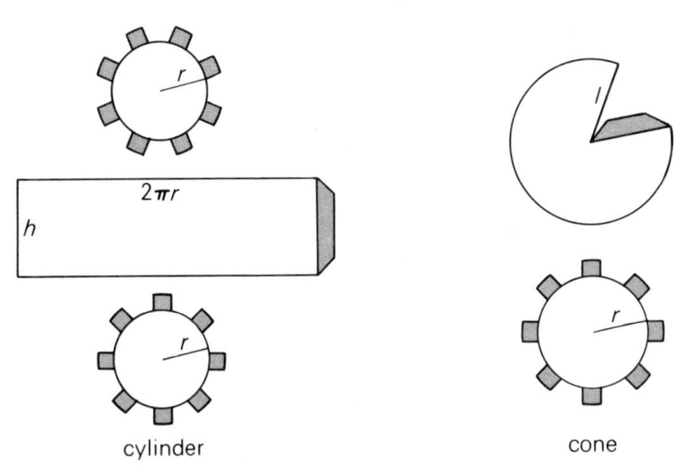

Notice that the net of the curved surface of a cylinder is a rectangle. The net of the curved surface of a cone is a sector of a circle whose radius is equal to the slant height of the cone.

Exercise 30a

(1) Draw the net of a cube on thin card. Cut out your net, leaving tabs, and from your net make a cube.

(2) Draw an equilateral triangle of side 6 cm on thin card, and join the mid-points of the sides to form the net of a tetrahedron. Cut out your net leaving tabs, then fold and glue to form a regular tetrahedron.

(3) Draw the net of an octahedron on thin card and so make a regular octahedron.

(4) Make a regular icosahedron out of thin card.

(5) On stiff card, draw a regular pentagon in a circle radius 3 cm, (see chapter 26). Cut out your pentagon and by drawing round it, draw the net of a dodecahedron. Cut out your net leaving tabs and from the net, make a regular dodecahedron.

(6) Copy and complete the following table:

solid	No. of faces	No. of vertices	No. of edges
tetrahedron			
cube			
octahedron			
dodecahedron			
icosahedron			

(7) Do the regular solids fit the formula $F + V - E = 2$?

(8) Which two regular solids have 12 edges? What do you notice about the values of F and V for these solids?

(9) Which two regular solids have 30 edges? What do you notice about the values of F and V for these solids?

(10) Which regular solid has 6 edges? What do you notice about the values of F and V for this solid?

(11) Draw the net of a square pyramid on thin card. Let the base have side 4 cm and let the sloping edges be 6 cm. Cut out the net and make a pyramid.

(12) Make a cone as follows: on thin card draw a circle with radius 4 cm and cut it out. Cut out a quadrant of the circle (90° sector) but leave a tab on the large sector as shown on page 143. Bend the large sector and glue. Draw and cut out a circle radius 3 cm, but leave tabs as shown on page 143. Bend the tabs and glue to form the base of the cone.

(13) Make a pair of regular tetrahedra with 3 cm edges. Glue them together to form a solid with six faces.

(14) Draw the net of cuboid with dimensions 4 cm $\times$ 2 cm $\times$ 1 cm. Cut out your net leaving tabs. Fold the net to make the cuboid.

(15) Make the curved surface of a cylinder from a rectangle of card 12 cm $\times$ 6 cm with tabs. Make an open cylinder from your net. Make a cylinder with a different radius from another rectangle 12 cm $\times$ 6 cm. Have the two cylinders got the same volume?

Exercise 30b

(1) (a) How many faces has an octahedron?

(b) If an octahedron has 6 vertices, use the relationship $F + V = E + 2$ to find the number of edges.

If each face of an octahedron is an equilateral triangle with side 3 cm and area 3.9 cm^2, calculate

(c) the total length of the edges

(d) the total surface area.

(2) (a) How many faces has an icosahedron?

(b) If an icosahedron has 12 vertices, calculate the number of edges.

(c) If each face of an icosahedron is an equilateral triangle of side 2 cm and height 1.7 cm, calculate the surface area of the icosahedron.

(3) (a) Which regular solid has 12 faces?

(b) What is the shape of each face of this solid?

(c) Which regular solid has 6 faces and 8 vertices?

(d) Which regular solid has 8 faces and 6 vertices?

(4) A cube has a surface area of 24 cm^2. Calculate

(a) the length of the side of the cube

(b) the volume of the cube.

If the length of side of the cube is doubled, calculate

(c) the new surface area

(d) the new volume.

(5) (a) Name a solid with 4 plane faces.

(b) Name a solid with 1 plane face, 1 curved surface, 1 edge and 1 vertex.

(c) Name a solid with 1 plane face, 1 curved surface, 1 edge and no vertex.

(d) Name a solid with no edges or vertices.

31 Straight Line Graph

Graphs in Algebra

A straight line graph which passes through the origin has an equation of the form $y = mx$, where m can have any value. The slope of the line depends on the value of m. The larger the value of m, the greater the slope of the line.

If you make up a table of values for the graph of $y = x$, and then add 2 to every value, you will have a new set of values which fit the equation $y = x + 2$.

x	-5	-4	-3	-2	-1	0	1	2	3	4	5
$y = x + 2$	-3	-2	-1	0	1	2	3	4	5	6	7

Adding 2 to each value moves the graph 2 units upwards. If you subtract 1 from each value of a function, it will move the graph 1 unit down.

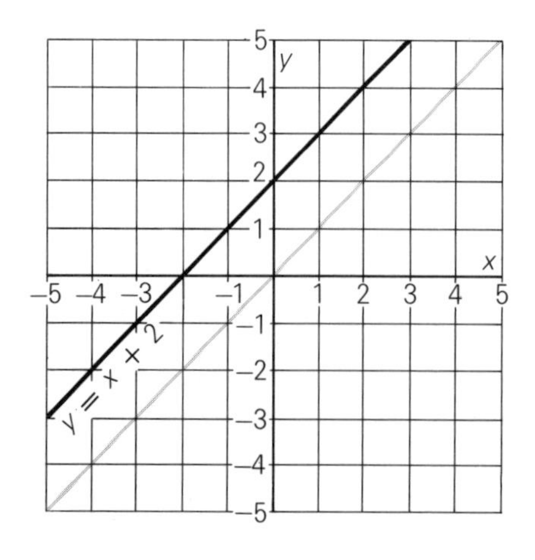

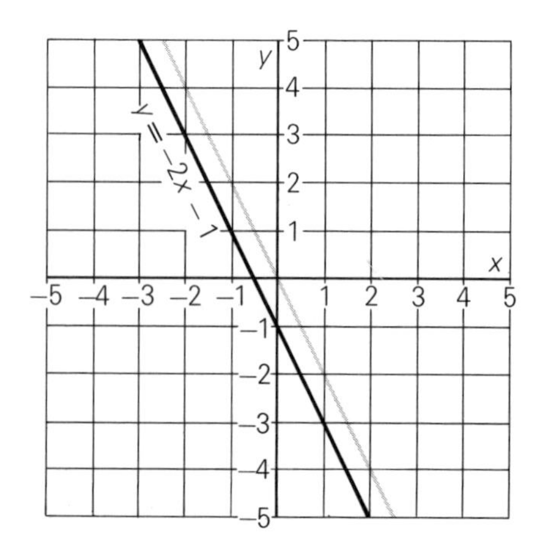

Graph of $y = x + 2$

Graph of $y = -2x - 1$

Slope of a Line

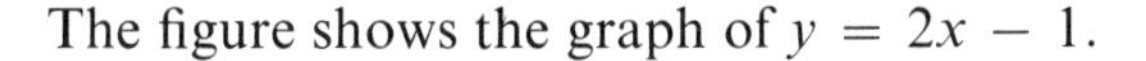
The figure shows the graph of $y = 2x - 1$.

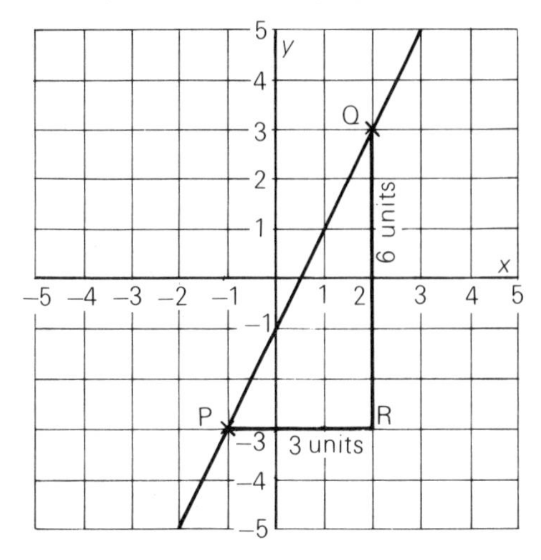

Graph of $y = 2x - 1$

To find the slope of the line, choose any two points P and Q on the line. Complete $\triangle$PQR by drawing PR parallel to the x-axis and QR parallel to the y-axis. The slope of the line PQ is

$$\frac{QR}{PR} = \frac{6 \text{ units}}{3 \text{ units}} = 2.$$

The slope of the line is also called the *gradient* of the line.

When finding the distances QR and PR, it is the number of units which is important, not the number of cm on the graph paper. Although $\triangle$PQR can be drawn on any part of the line, the calculation is easier if PR is a whole number of units.

For lines such as the graph of $y = -2x - 1$ on page 146, which slopes the other way, the slope is found in the same way as before but the distance from P to R is in the negative direction, so the slope is negative.

Conversion Graphs

Conversion graphs are drawn to change from one unit to another, or as a kind of ready reckoner. If we know two points on the graph we can plot them and draw a straight line through them to complete the graph. The graph shown overleaf is for changing £ to dollars. The graph can be drawn if we know that 2.35 dollars can be bought for £1. (We know that £0 buys 0 dollars and £10 buys 23.5 dollars.)

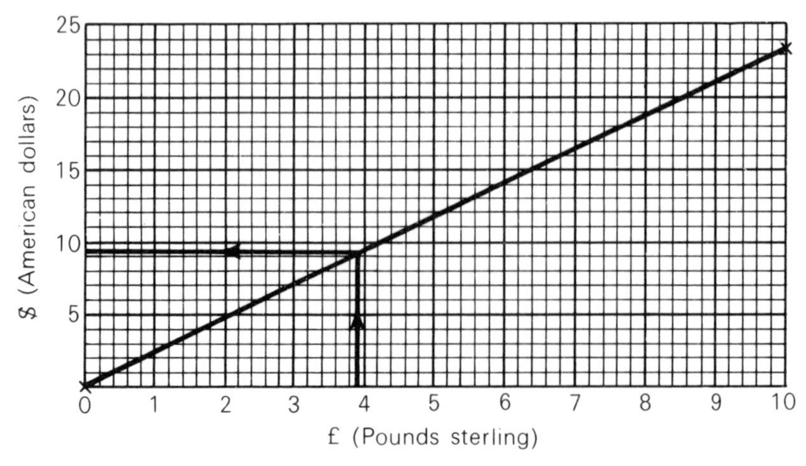

We can see from the graph that £3.90 buys just over \$9, but the graph is not very accurate because the scale is not large enough.

Exercise 31a

(1) Copy and complete the table of values for the graph of $y = 2x - 1$.

x	-5	-4	-3	-2	-1	0	1	2	3	4	5
$2x$	-10	-8		-4	-2		2	4	6		10
-1		-1	-1	-1		-1	-1	-1		-1	-1
y	-11		-7	-5	-3	-1	1			7	9

Draw the graph using a scale of 1 cm to 1 unit on both axes.

(2) Make a table of values for the graph of $y = x + 3$ from $x = -4$ to $x = +4$. Draw the graph using a scale of 1 cm to 1 unit on both axes.

(3) Make a table of values for the graph of $y = 4x - 3$ from $x = -1$ to $x = +4$. Draw the graph using a scale of 2 cm to 1 unit of x and 1 cm to 1 unit of y.

(4) Make a table of values for the graph of $y = 3x + 5$ from $x = -5$ to $x = +5$. Draw the graph using a scale of 1 cm to 1 unit of x and 2 cm to 5 units of y.

(5) Draw the graph of $y = 2x - 3$, and use the method shown on page 147 to find the slope of the line.

(6) Draw the graph of $y = \frac{1}{2}x + 3$ and find the slope of the graph.

(7) Draw axes and plot the points $(2, 3)$ and $(4, 6)$. Draw a straight line through the points and find the slope of the line.

(8) Plot the points $(-2, 1)$ and $(1, 10)$. Find the slope of the line through these points.

(9) Find the slope of the line which passes through the points (2, 7) and (5, 1).
(10) Find the slope of the line which passes through the points $(-1, 5)$ and $(2, -1)$.
(11) Draw a graph to convert £ to German Marks, for amounts up to £10 if £1 = DM5.50.
(12) Draw a graph to convert degrees Fahrenheit to degrees Celsius, given that 0°C = 32°F and 100°C = 212°F.

Exercise 31b

(1) Make a table of values for the graph of $y = x + 5$ from $x = -4$ to $x = +4$, and

(a) plot the graph using a scale of 2 cm to 1 unit on both axes,

(b) from your graph find the value of (i) y when $x = -2$ (ii) x when $y = 7$,

(c) find the slope of the graph.

(2) Make a table of values for the graph of $y = -2x + 3$ from $x = -4$ to $x = +4$, and

(a) plot the graph using a scale of 1 cm to 1 unit on both axes,

(b) from your graph find the value of (i) y when $x = 3\frac{1}{2}$ (ii) x when $y = -2$,

(c) find the slope of the graph.

(3) Make a table of values for the graph of $y = 2x - 1$ from $x = -3$ to $+3$, and

(a) plot the graph using a scale of 2 cm to 1 unit of x and 1 cm to 1 unit of y,

(b) find the slope of your graph,

(c) on the same axes draw the line which passes through the points $(-1, 6)$ and $(3, 2)$,

(d) find the points of intersection of the two lines.

(4) (a) Draw the graph of $y = 3x - 3$ from $x = -3$ to $x = +3$ using any suitable scale.

(b) Calculate the gradient of your line.

(c) Write down the co-ordinates of the point where the graph cuts the x-axis.

(d) Calculate the area of the triangle enclosed by the graph and the axes.

32 Pythagoras

The Pythagoras Rule

If you know the lengths of two sides of a right-angled triangle, the Pythagoras rule will help you to find the third side. The rule says that in any right-angled triangle, the square of the hypotenuse is equal to the sum of the squares of the other two sides.

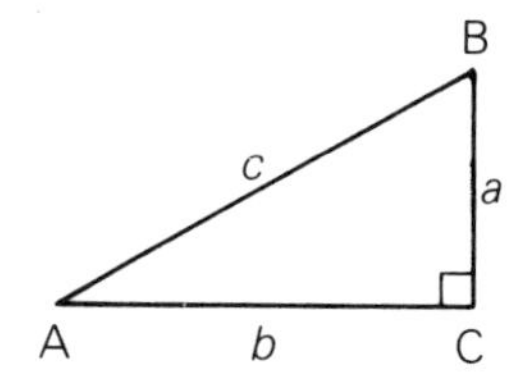

In the figure, this means that

$a^2 + b^2 = c^2$.

Thus, if the two shorter sides are known, the hypotenuse can be found. Also, by subtracting b^2 from both sides, we get

$a^2 = c^2 - b^2$

which can be used if the hypotenuse and one other side are known.

Using the Rule

The rule can be used whenever there is a right-angled triangle. It may be that some or all of the sides are a whole number of units, but you will probably need to use square root tables in order to find the third side.

Example: A telegraph pole is held in position by steel cables, each of which is joined to the top of the pole and to a point on the ground which is 8 m from the base of the pole. If the cables are straight and 11 m long, find the height of the pole, correct to 2 significant figures.

Let the height of the pole be h metres.

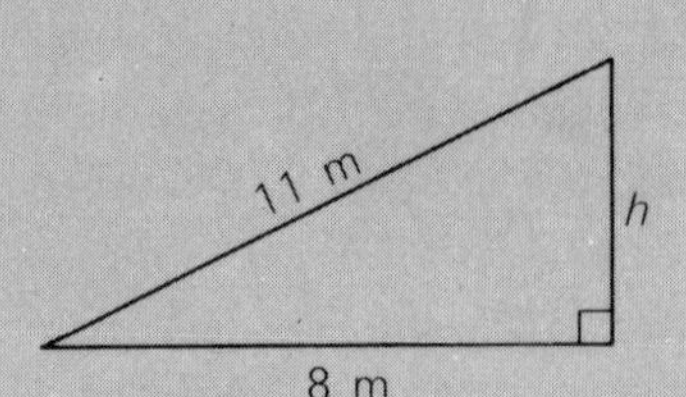

$$\begin{aligned} \text{Then } h^2 &= 11^2 - 8^2 \\ &= 121 - 64 \\ &= 57 \\ \therefore h &= \sqrt{57} \\ &= 7.550 \text{ (from tables)} \\ &= 7.6 \text{ m (correct to 2 significant figures).} \end{aligned}$$

Exercise 32a

Questions 1–10 refer to △ABC shown on page 150. In each case calculate the unknown side, correct to 3 significant figures.

(1) $a = 3$ cm and $b = 8$ cm
(2) $a = 7$ cm and $b = 9$ cm
(3) $a = 6$ m and $b = 11$ m
(4) $a = 3$ km and $b = 12$ km
(5) $a = 15$ km and $c = 20$ km
(6) $a = 7$ cm and $c = 12$ cm
(7) $c = 11$ cm and $a = 3$ cm
(8) $b = 9$ m and $c = 13$ m
(9) $c = 25$ m and $b = 16$ m
(10) $c = 30$ cm and $b = 20$ cm.

(11) Calculate the diagonal of a rectangle if the sides are 6 cm and 10 cm.
(12) Calculate the diagonal of a rectangle whose sides are 9 cm and 15 cm.
(13) Calculate the width of a rectangle if the length is 14 cm and the diagonal is 17 cm.
(14) Calculate the length of a rectangle with width 5 cm and diagonal 11 cm.
(15) Calculate the diagonal of a square of side 3 m.
(16) Calculate the diagonal of a square of side 15 cm.
(17) Calculate the two equal sides of an isosceles triangle with base 10 cm and height 7 cm.
(18) Calculate the height of an isosceles triangle with sides 12 cm, 12 cm and 8 cm.
(19) Calculate the height of an equilateral triangle with sides 6 cm long.

(20) Calculate the distance between the parallel sides of an isosceles trapezium if the equal sides are 3 cm and the parallel sides are 6 cm and 8 cm.

Exercise 32b

(1) The diagonals of a rhombus are 6 cm and 10 cm.
(a) Calculate the length of the sides of the rhombus.
(b) Draw the rhombus as accurately as possible.
(c) Measure the length of the sides correct to the nearest mm.
(d) Calculate the area of the rhombus.

(2) ABCD is a kite whose diagonals intersect at X. If AX = XC = 7 cm, BX = 3 cm and XD = 9 cm.
(a) Calculate the length of AB.
(b) Calculate the length of AD.
(c) Make an accurate drawing of the kite and measure $\angle$ABC correct to the nearest degree.
(d) Calculate the area of ABCD.

(3) A chord AB of a circle radius 6 cm, is 4 cm from the centre O.
(a) Calculate the length of the chord.
(b) Calculate the area of $\triangle$AOB.
(c) Make an accurate drawing of the circle and the chord.
(d) Measure AB correct to the nearest mm, and $\angle$AOB correct to the nearest degree.

(4) P is a point 10 cm from the centre O of a circle radius 5 cm. Tangents AP and BP are drawn from P to the circle.
(a) Calculate the length of the tangents.
(b) Make an accurate drawing of the circle and tangents.
(c) Measure the length of the chord of contact AB, correct to the nearest mm.
(d) Measure $\angle$APB, correct to the nearest degree.

(5) ABCD is a square of side 7 cm.
(a) Calculate the area of ABCD.
(b) Calculate the length of AC.
(c) Calculate the area of the circle with AB as diameter.
(d) Calculate the area of the circle with AC as radius.
In (c) and (d) take $\pi = \frac{22}{7}$.

Section F
33 Simple Trigonometry

Sine, Cosine, Tangent

If $\triangle ABC$ has a right angle at C as shown below, the *tangent* of $\angle A$ or $\angle B$ is

$$\frac{\text{side opposite the angle}}{\text{side next to the angle}}.$$

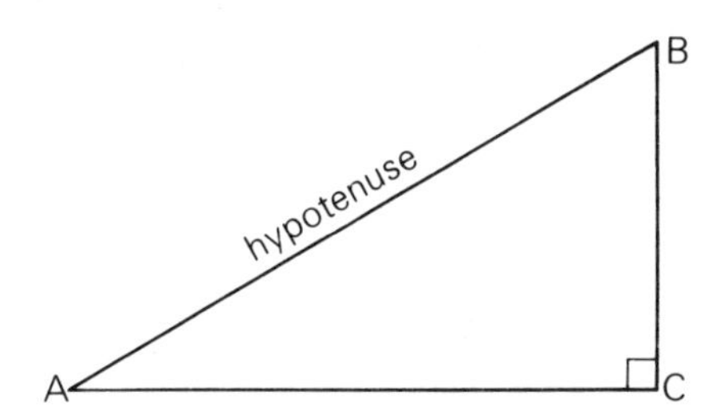

Thus, $\tan A = \dfrac{BC}{AC}$ and $\tan B = \dfrac{AC}{BC}$.

The *sine* of $\angle A$ or $\angle B$ is

$$\frac{\text{side opposite the angle}}{\text{hypotenuse}}.$$

Thus, $\sin A = \dfrac{BC}{AB}$ and $\sin B = \dfrac{AC}{AB}$.

The *cosine* of $\angle A$ or $\angle B$ is

$$\frac{\text{side next to the angle}}{\text{hypotenuse}}.$$

Thus, $\cos A = \dfrac{AC}{AB}$ and $\cos B = \dfrac{BC}{AB}$.

Four-Figure Tables

Four-figure tables give the sine, cosine and tangent of angles from 0° to 90°. Each degree is divided into 60 minutes. (1° = 60′.)

Using sine, cosine, or tangent tables is something like using log tables.

Tangent Tables

The following programme shows how to use tangent tables.

NATURAL TANGENTS

	0′	10′	20′	30′	40′	50′	Differences 1′	2′	3′	4′	5′	6′	7′	8′	9′
0°	0.0000	0.0029	0.0058	0.0087	0.0116	0.0145	3	6	9	12	15	17	20	23	26
1	0.0175	0.0204	0.0233	0.0262	0.0291	0.0320	3	6	9	12	15	17	20	23	26
2	0.0349	0.0378	0.0407	0.0437	0.0466	0.0495	3	6	9	12	15	18	20	23	26
3	0.0524	0.0553	0.0582	0.0612	0.0641	0.0670	3	6	9	12	15	18	20	23	26
4	0.0699	0.0729	0.0758	0.0787	0.0816	0.0846	3	6	9	12	15	18	21	23	26
5	0.0875	0.0904	0.0934	0.0963	0.0992	0.1022	3	6	9	12	15	18	21	24	26
6	0.1051	0.1080	0.1110	0.1139	0.1169	0.1198	3	6	9	12	15	18	21	24	27
7	0.1228	0.1257	0.1287	0.1317	0.1346	0.1376	3	6	9	12	15	18	21	24	27
8	0.1405	0.1435	0.1465	0.1495	0.1524	0.1554	3	6	9	12	15	18	21	24	27
9	0.1584	0.1614	0.1644	0.1673	0.1703	0.1733	3	6	9	12	15	18	21	24	27
10°	0.1763	0.1793	0.1823	0.1853	0.1883	0.1914	3	6	9	12	15	18	21	24	27

Answers ↓	
	Look at the tangent table above. Move your finger down the first column to 8°. The number next to it has a ring round it. What number is it?
0.1405	0.1405 is in the second column. What is the heading of this column?
0′	0.1405 = tan 8° 0′. Move your finger across to the number in the 30′ column. This also has a ring round it. it is tan 8° 30′. What is the value of tan 8° 30′?
0.1495	What is the value of tan 5° 50′?
0.1022	Open your own tables at the page headed 'Natural Tangents'. Use your tables to find tan 25° 20′.

0.4734	What is the value of tan 58° 40′?
1.6426	Now look at page 154, again. Move your finger down the page to 8° and across to the tangent of 8° 30′. What is its value?
0.1495	To find tan 8° 33′ we must add the number in the 3 difference column. Keep your finger at 0.1495 and move another finger across to the 3 difference column. What number do you find?
9	What do you get if you 'add' 9 to 0.1495?
0.1504	This shows that tan 8° 33′ = 0.1504. Now look at the tangent tables in your own book. What is the value of tan 18° 27′?
0.3337	What is the value of tan 53° 51′?
1.3688	What is the value of tan 41° 11′?
0.8749	Now do Exercise 33a, questions 1–10, and then go on to the next part of the programme.

Answers ↓	Suppose we know that the tangent of an angle is 0.0407 and we want to find the angle. In this case we must look for 0.0407 in the tables. Look on page 154 again. 0.0407 has a ring round it. What angle is it the tangent of?
2° 20′	Notice that the larger the angle, the larger the tangent. Turn to your own tangent tables now. Can you find 0.3772 in the tables? What angle is it the tangent of?
20° 40′	Now look for 5.976. What angle is it the tangent of?
80° 30′	What angle is 2.211 the tangent of?
65° 40′	2.194 = tan 65° 30′. Can you find 2.196 in the tables?
No	2.196 is just greater than 2.194. What is the 'difference' between 2.196 and 2.194?
2	Move your finger across the 65° line. Which difference column has a 2 in it?
The first	What is the heading of the first difference column?

1′	This shows that the angle whose tangent is 2.196 is 1′ more than the angle whose tangent is 2.194 or 65° 30′ + 1′ = 65° 31′. Which angle has a tangent of 2.196?
65° 31′	Which angle has a tangent of 0.5206?
27° 30′	Which angle has a tangent of 0.5210?
27° 31′	Which angle has a tangent of 0.5221?
27° 34′	Which angle has a tangent of 1.1303?
48° 30′	Which angle has a tangent of 1.1323?
48° 33′	Which angle has a tangent of 0.8026?
38° 45′	Which angle has a tangent of 0.2720?
15° 13′	Now do Exercise 33a, questions 11–20.

Sine Tables

Sine tables are used in exactly the same way as tangent tables. If you look at your sine tables you will see that there is no whole number in front of the decimal point in any part of the tables except at 89°. This is because the number before the point is zero at all parts of the table. Before you go on to the cosine programme, make sure that you can do Exercise 33a, questions 21–40.

Cosine Tables

Answers ↓	
	Open your book of tables to the page of cosines. What is the heading at the top of the page?
Natural cosines	To start with use these tables the same way that you would use sine or tangent tables. What is the value of cos 12°?
0.9781	Notice that if there is no whole number you must put an 0 in front of the decimal point. What is the value of cos 61° in your tables?

0.4848	Notice that as the angle gets larger the cosine gets smaller. What is the value of cos 35° 40′?
0.8124	What is the value of cos 72° 20′?
0.3035	Now look at the cosine table below. What is the value of cos 7° 20′?
0.9918	Suppose we want to find cos 7° 22′. Keep your finger at 0.9918 and move another finger across to the 2′ difference column. What number do you find?
1	Because a larger angle means a smaller cosine, the 1 must be subtracted from 0.9918. What is the value of cos 7° 22′?
0.9917	Turn back to your own tables. What is the value of cos 63° 44′? (Remember to subtract the number in the difference column.)
0.4426	What is the value of cos 19° 37′?
0.9419	What is the value of cos 84° 41′?
0.0926	Now do Exercise 33a, questions 41–50, then go on with the programme.

NATURAL COSINES

	0′	10′	20′	30′	40′	50′	Differences to be subtracted								
							1′	2′	3′	4′	5′	6′	7′	8′	9′
0°	1.0000	1.0000	1.0000	1.0000	.9999	.9999	0	0	0	0	0	0	0	0	0
1	.9998	.9998	.9997	.9997	.9996	.9995	0	0	0	0	0	0	0	0	0
2	.9994	.9993	.9992	.9990	.9989	.9988	0	0	0	0	0	0	1	1	1
3	.9986	.9985	.9983	.9981	.9980	.9978	0	0	0	1	1	1	1	1	2
4	.9976	.9974	.9971	.9969	.9967	.9964	0	0	1	1	1	1	2	2	2
5	.9962	.9959	.9957	.9954	.9951	.9948	0	1	1	1	1	2	2	2	2
6	.9945	.9942	.9939	.9936	.9932	.9929	0	1	1	1	2	2	2	3	3
7	.9925	.9922	.9918	.9914	.9911	.9907	0	1	1	2	2	2	3	3	3
8	.9903	.9899	.9894	.9890	.9886	.9881	0	1	1	2	2	3	3	3	4
9	.9877	.9872	.9868	.9863	.9858	.9853	0	1	1	2	2	3	3	4	4
10°	.9848	.9843	.9838	.9833	.9827	.9822	1	1	2	2	3	3	4	4	5
11	.9816	.9811	.9805	.9799	.9793	.9787	1	1	2	2	3	3	4	5	5
12	9781	.9775	.9769	.9763	.9757	9750	1	1	2	3	3	4	4	5	6
13	.9744	.9737	.9730	.9724	.9717	.9710	1	1	2	3	3	4	5	5	6
14	.9703	.9696	.9689	.9681	.9674	.9667	1	1	2	3	4	4	5	6	7
15	.9659	.9652	.9644	.9636	.9628	.9621	1	2	2	3	4	5	5	6	7

Answers ↓	
	Look at page 157, and find 0.9989 in the table. It has a ring round it. Which angle has a cosine of 0.9989?
2° 40′	Turn to your own tables. The larger the angle, the smaller the cosine. Which angle has a cosine of 0.1679?
80° 20′	Which angle has a cosine of 0.6626?
48° 30′	Which angle has a cosine of 0.9013?
25° 40′	Now turn back to the table on page 157. Which angle has a cosine of 0.9750?
12° 50′	The larger the angle the smaller the cosine. If 0.9750 = cos 12° 50′, is 0.9745 the cosine of a larger angle or a smaller angle?
larger	0.9750 = cos 12° 50′. What is the ‘difference’ between 0.9750 and 0.9745?
5	Look along the 12° row. In which difference column is there a 5?
8′	Because 0.9745 is 5 less than 0.9750, it is the cosine of an angle 8′ larger than 12° 50′. Which angle has a cosine of 0.9745?
12° 58′	Now turn to your own cosine tables for the angle whose cosine is 0.2664. Which number in the table is just larger than 0.2664?
0.2672	Which angle has a cosine of 0.2672?
74° 30′	What is the ‘difference’ between 0.2672 and 0.2664?
8	Move your finger across the 74° row until you get to an 8 in the difference column. In which difference column is there an 8?
3′	The 3′ must be added to 74° 30′. Which angle has a cosine of 0.2664?
74° 33′	Which angle has a cosine of 0.5983?
53° 15′	Which angle has a cosine of 0.9230?
22° 38′	Now do Exercise 33a, questions 51–60.

Example: In $\triangle ABC$, $AC = 7$ cm, $BC = 12$ cm and $\angle C = 90°$. Calculate $\angle A$ correct to the nearest minute.

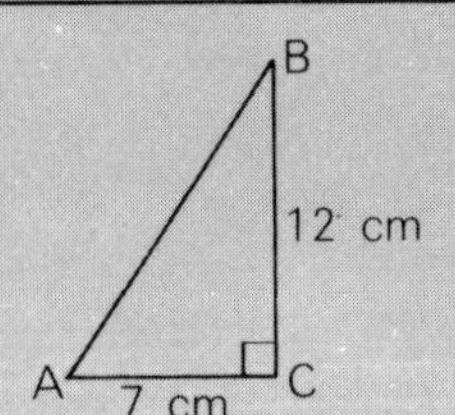

$$\tan A = \frac{\text{opposite}}{\text{adjacent}} = \frac{BC}{AC} = \frac{12}{7}$$

$= 1.7143$ (correct to 4 decimal places).

$\therefore \angle A = 59° 45'$ (from the tables).

Exercise 33a

Write down the tangent of each of the following:

(1) 23° 30′ (2) 56° 20′ (3) 39° 10′ (4) 47° 32′ (5) 63° 14′ (6) 77° 27′ (7) 18° 4′ (8) 29° 11′ (9) 4° 51′ (10) 66° 53′.

Write down the angle whose tangent is:

(11) 0.7089 (12) 2.496 (13) 0.3378 (14) 0.7729 (15) 0.2953 (16) 0.3902 (17) 0.2883 (18) 0.5638 (19) 1.2476 (20) 0.4938

Write down the sine of each of the following:

(21) 23° 10′ (22) 14° 20′ (23) 81° 20′ (24) 43° 18′ (25) 21° 36′ (26) 42° 19′ (27) 57° 21′ (28) 48° 13′ (29) 9° 43′ (30) 61° 23′.

Write down the angle whose sine is:

(31) 0.3907 (32) 0.6713 (33) 0.9239 (34) 0.4478 (35) 0.9316 (36) 0.6132 (37) 0.9756 (38) 0.2874 (39) 0.2289 (40) 0.6611

Write down the cosine of each of the following:

(41) 61° 20′ (42) 43° 30′ (43) 27° 10′ (44) 39° 51′ (45) 64° 26′ (46) 83° 7′ (47) 23° 14′ (48) 18° 54′ (49) 46° 27′ (50) 38° 26′.

Write down the angle whose cosine is:

(51) 0.9367	(54) 0.2885	(57) 0.4710	(60) 0.7248.
(52) 0.7274	(55) 0.8961	(58) 0.8459	
(53) 0.4358	(56) 0.5790	(59) 0.4794	

Exercise 33b

(1) In $\triangle ABC$, AC = 5 cm, BC = 17 cm and $\angle C = 90°$.
(a) Calculate tan A.
(b) Use tangent tables to find $\angle A$, correct to the nearest minute.
(c) Calculate AB, correct to the nearest mm.
(d) Calculate the area of $\triangle ABC$.

(2) In $\triangle PQR$, PR = 5.34 cm, QR = 2.17 cm and $\angle R = 90°$.
(a) Use logarithms to calculate tan Q, correct to 3 significant figures.
(b) Use your answer to (a) to find $\angle Q$, correct to the nearest degree.
(c) Calculate $\angle P$, correct to the nearest degree.
(d) Use logarithms to calculate the area of $\triangle PQR$ in cm^2, correct to 3 significant figures.

(3) In $\triangle DEF$, DE = 7 cm, EF = 5 cm and $\angle F = 90°$.
(a) Calculate sin D correct to 4 decimal places.
(b) Use tables to find $\angle D$ correct to the nearest minute.
(c) Find $\angle E$ correct to the nearest minute.
(d) Calculate DF correct to the nearest mm.

(4) A rhombus PQRS has PR = 8 cm and PQ = 12 cm. Calculate
(a) the cosine of $\angle QPR$ correct to 4 decimal places
(b) $\angle QPR$
(c) QS correct to the nearest mm
(d) $\angle PQR$.

(5) The diagonals of a kite ABCD intersect at X. If AX = XC = 12 cm, BX = 5 cm and AD = 16 cm, calculate
(a) AB
(b) the tangent of $\angle BAC$ and hence find $\angle BAC$
(c) the cosine of $\angle ABD$ and hence find $\angle ABD$
(d) the sine of $\angle BDA$ and hence find $\angle BDA$.

34 Finite Number Systems

Modulo Arithmetic

Because $26 \div 3 = 8$ remainder 2, we say that 26 is *congruent* to 2 in *modulo* 3, and we write

$26 \equiv 2 \pmod 3$.

Here is the set of natural numbers. It is an infinite set.

$N = \{1, \mathbf{2}, 3, 4, \mathbf{5}, 6, 7, \mathbf{8}, 9, 10, \ldots\}$.

The numbers in heavy type all give a remainder of 2 when they are divided by 3. They are all congruent to 2 (mod 3).

Here is the set of numbers which is congruent to 2 (mod 3). It is also an infinite set.

$P = \{2, 5, 8, 11, 14, 17, \ldots\}$.

Because these numbers are congruent to 2 they are shown by a 2 in square brackets.

$[2]_3 = \{2, 5, 8, 11, 14, 17, \ldots\}$.

The little 3 shows that we are working in modulo 3. In the same way

$[1]_3 = \{1, 4, 7, 10, 13, 16, \ldots\}$
$[3]_5 = \{3, 8, 13, 18, \ldots\}$.

If we work in modulo 5, we can say that $4 + 3 = 7 \equiv 2 \pmod 5$. Another way of writing this is

$4 \oplus_5 3 = 2$.

Using the $\oplus_5$ sign, means 'add the two numbers, divide by 5 and take the remainder'. In this way, we can make an addition or multiplication table for modulo 5.

$\oplus_5$	1	2	3	4	0
1	2	3	4	0	1
2	3	4	0	1	2
3	4	0	1	2	3
4	0	1	2	3	4
0	1	2	3	4	0

$\otimes_5$	1	2	3	4	0
1	1	2	3	4	0
2	2	4	1	3	0
3	3	1	4	2	0
4	4	3	2	1	0
0	0	0	0	0	0

Notice that the addition and multiplication tables for mod 5 are symmetrical about the dotted line. Why is this? What does it mean?

Bases and Remainders

Let us see what happens when we change numbers from base 10 to modulo 10. Take any three numbers.

$46_{10} \equiv 6 \pmod{10}$
$39_{10} \equiv 9 \pmod{10}$
$73_{10} \equiv 3 \pmod{10}$.

In each case the remainder when we divide by 10 is the units digit of the number in base 10. This is because the modulo number is the same as the base number. Does this also work for other bases and modulo numbers? Try base 7.

$35_7 = 26_{10} \equiv 5 \pmod{7}$.

Again, we see that the remainder when we divide by 7 is the same as the units digit of the number in base 7. Does this work for all bases and modulo numbers? If so, why?

Digital Roots

If you add the digits of a number the result is called the *digital root*.

Example: Find the digital root of 624.

624 → [$6 + 2 + 4 = 12$; $1 + 2 = 3$] → 3

For any two numbers, the sum of the digital roots is equal to the digital root of the sum. For example, $463 + 591 = 1\,054$. The digital root of 463 is 4 and of 591 is 6. Adding these gives $4 + 6 = 10$. The digital root of 10 and of 1 054 is 1.

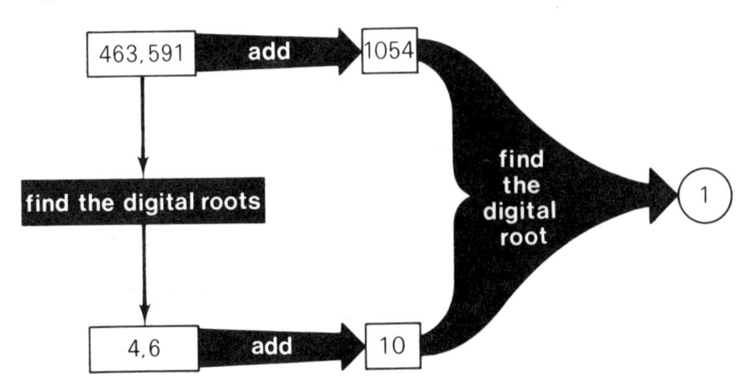

Now change 'add' to multiply. Does it still work? If so, why?

Exercise 34a

(1) List the first 5 elements of $[4]_5$.
(2) List the first 5 elements of $[0]_4$.
(3) List the first 5 elements of $[1]_2$.
(4) List the first 5 elements of $[2]_{10}$.

Say whether the following statements are true or false:

(5) $3 \oplus_8 7 = 2$
(6) $5 \oplus_9 3 = 8$
(7) $3 \oplus_8 6 = 5$
(8) $4 \oplus_5 2 = 2$
(9) $2 \oplus_6 4 = 0$
(10) $0 \oplus_6 4 = 2$
(11) $3 \otimes_5 3 = 4$
(12) $4 \otimes_6 5 = 0$
(13) $2 \otimes_3 2 = 2$
(14) $3 \otimes_6 3 = 3$
(15) $6 \otimes_9 4 = 1$
(16) $4 \otimes_6 4 = 4$
(17) $32_{10} \equiv 2 \pmod 5$
(18) $32_{10} \equiv 2 \pmod 6$
(19) $32_{10} \equiv 2 \pmod{10}$
(20) $51_{10} \equiv 5 \pmod 4$
(21) $36_{10} \equiv 3 \pmod 7$
(22) $41_{10} \equiv 7 \pmod{11}$
(23) $63_{10} \equiv 3 \pmod 5$
(24) $49_{10} \equiv 0 \pmod 7$
(25) $72_{10} \equiv 3 \pmod 9$
(26) $63_{10} \equiv 3 \pmod{10}$
(27) $42_6 \equiv 2 \pmod 6$
(28) $38_7 \equiv 8 \pmod{10}$
(29) $61_6 \equiv 7 \pmod 6$
(30) $84_9 \equiv 4 \pmod 9$.

Find the digital root of the following numbers:

(31) 46 (32) 68 (33) 29 (34) 37 (35) 64 (36) 563 (37) 428 (38) 817 (39) 409 (40) 623.

(41) Add the digital roots of 673 and 412. Check by finding the digital root of (673 + 412).
(42) Add the digital roots of 314 and 206. Check by finding the digital root of (314 + 206).
(43) Add the digital roots of 417 and 216. Check by finding the digital root of (417 + 216).
(44) Multiply the digital roots of 28 and 42. Check by finding the digital root of 28 × 42.
(45) Multiply the digital roots of 16 and 77. Check by finding the digital root of 16 × 72.

Exercise 34b

(1) (a) Make a table for addition in modulo 3.
(b) If $35 \equiv n \pmod 3$ write down the value of n.
(c) Change 28 to modulo 3 and add the value of n in (b). Check by changing (35 + 28) to modulo 3.
(d) List the first 5 elements of $[2]_3$.
(2) (a) Make a table for addition in modulo 4.
(b) Make a table for multiplication in modulo 4.
(c) If $n^2 = 0$, and $n \neq 0$, use your table to find the value of n.
(d) Calculate $23 \otimes_4 31$.
(3) (a) Copy and complete the addition table for modulo 2.

$\oplus_2$	1	0
1		1
0		

$\otimes_2$	1	0
1		0
0		

(b) Copy and complete the multiplication table for modulo 2.
(c) List the first 5 elements of $[0]_2$.
(d) Describe the set $[1]_2$.
(4) (a) List the first five elements of $[3]_6$.
(b) Calculate $321 \otimes_6 44$.
(c) Calculate $27 \oplus_6 39$.
(d) List the first five elements of $[5]_6$. What kind of numbers are they? Find an element of $[1]_6$ which is not a prime number.
(5) (a) Find the digital root of 462 731.
(b) Find the digital root of 43 625 + 517 031.
(c) Find the digital root of 14 896 × 30 425.
(d) Calculate $476 \otimes_9 29$.

35 Logarithms

Squares

Often, when using logs we need the square of a number. There are two ways of finding a square. You could either write the log of the number twice and add, or you could multiply the log by 2.

Example: Calculate $(29.1)^2$, correct to 3 significant figures.

Method 1

No	Log
29.1	1.4639
29.1	1.4639
846.8	2.9278

$(29.1)^2 = 847$ (correct to 3 significant figures).

Method 2

No	Log
29.1	1.4639
	× 2
846.8	2.9278

$(29.1)^2 = 847$ (correct to 3 significant figures).

Square Roots

To find the square of a number, its log is multiplied by 2. To find the square root of a number its log is divided by 2.

Example: Calculate $\sqrt{35.6}$, correct to 3 significant figures.

No	Log
35.6	1.5514
	÷ 2
5.967	0.7757

$\sqrt{35.6} = 5.97$ (correct to 3 significant figures).

If you need to find the square root of a number less than 1, it is easier to use square root tables.

Finding the square root of a number is often part of a larger calculation.

Example: Calculate the radius of a circle with area 50 cm² ($r = \sqrt{\left(\frac{A}{\pi}\right)}$, take $\pi = 3.142$).

No	Log
50	1.6990
3.142	0.4972
	1.2018
	÷ 2
3.989	0.6009

$r = 3.99$ cm (correct to 3 significant figures).

Cubes

When finding the cube of a number by logs, it is easier to multiply the log by 3 than to write it 3 times and add.

Exercise 35a

Use logarithms to calculate the following correct to 3 significant figures.

(1) $(3.61)^2$
(2) $(4.25)^2$
(3) $(12.7)^2$
(4) $(3.95)^2$
(5) $(0.735)^2$
(6) $(1.06)^3$
(7) $(2.31)^3$
(8) $(5.22)^3$
(9) $(2.17)^3$
(10) $(0.95)^3$
(11) $\sqrt{(3.76)}$
(12) $\sqrt{(45.8)}$
(13) $\sqrt{(61.3)}$
(14) $\sqrt{(84.2)}$
(15) $\sqrt{(8.64)}$
(16) $(4.733)^2$
(17) $(2.153)^3$
(18) $\sqrt{(42.16)}$
(19) $\sqrt{(4.975)}$
(20) $(2.245)^2$
(21) $(3.16 \times 4.05)^2$
(22) $(6.72 \times 1.63)^2$
(23) $(4.85 \times 2.76)^2$
(24) $4.16 \times (2.35)^2$
(25) $(3.62)^2 \times 4.16$
(26) $\sqrt{(3.71 \times 4.84)}$
(27) $\sqrt{(32.6 \times 4.78)}$
(28) $\sqrt{(6.27 \times 1.35)}$
(29) $\sqrt{(83.2 \div 4.29)}$
(30) $\sqrt{\frac{97.3}{23.6}}$.

Check your answers to questions 11–15 by using square root tables.

Exercise 35b

Use logarithms to evaluate the following, giving all your answers correct to 3 significant figures.

(1) A solid copper cube has edges 3.25 cm long. Calculate
  (a) the surface area of the cube
  (b) the volume of the cube
  (c) the weight of the cube, if copper weighs 8.94 g/cm³.

(2) A solid silver rod is in the shape of a cylinder of radius 2.35 cm, and length 12.1 cm. Calculate
  (a) the circumference of the rod ($C = 2\pi r$)
  (b) the area of cross-section ($A = \pi r^2$)
  (c) the volume of the rod ($V = \pi r^2 l$)
  (d) the weight of the rod, if silver weighs 10.5 g/cm³.

(3) A solid gold sphere has a radius of 1.07 cm. Calculate
  (a) the surface area of the sphere ($A = 4\pi r^2$)
  (b) the volume of the sphere ($V = \frac{4\pi r^3}{3}$)
  (c) the weight of the sphere if gold weighs 19.5 g/cm³.

(4) Three squares, each of side 2.75 cm are put side by side to form a rectangle ABCD.
  (a) Calculate the length of AC.
  (b) Calculate the area of the rectangle.
  (c) If PQRS is a square with the same area as ABCD, calculate the length of PQ.

(5) A circle has an area of 40 cm².
  (a) Calculate the radius of the circle if $r = \sqrt{\frac{A}{\pi}}$, (take $\pi = 3.142$).
  (b) If PQRS is a square with the same area as the circle, calculate the length of PQ.
  (c) Calculate the length d of the diagonal of PQRS if $d = \sqrt{(2A)}$.

36 Simultaneous Equations

Pairs of Equations

If a pair of equations, each with a term in x and a term in y, has a solution which fits both equations, then the equations are called *simultaneous equations*.

As with simple equations, it is important to do the same thing to both sides of the equation. Because the two equations are quite separate, it is not necessary to do the same to both equations.

Example: Solve the equations $\left.\begin{aligned} 3x - 2y &= 8 \\ x + 5y &= -3 \end{aligned}\right\}$

$$3x - 2y = 8 \;\ldots\ldots\ldots\ldots (1)$$
$$x + 5y = -3 \;\ldots\ldots\ldots (2)$$

Multiply (1) by 5

$$15x - 10y = 40 \;\ldots\ldots\ldots\ldots (3)$$

Multiply (2) by 2

$$2x + 10y = -6 \;\ldots\ldots\ldots (4)$$

Add (3) and (4)

$$17x = 34$$
$$x = 2.$$

Substitute $x = 2$ in (1)

$$\begin{aligned} 3 \times 2 - 2y &= 8 \\ 6 - 2y &= 8 \\ -2y &= 2 \\ -y &= 1 \\ y &= -1. \end{aligned}$$

Check in (2)
LHS $2 + 5(-1) = 2 - 5 = -3$. RHS -3.

Providing there are no fractions in the equations, they can be solved as in the example on page 168. The method is as follows:

1. Make the coefficient of y the same in both equations.
2. Add or subtract the equations to eliminate (get rid of) y.
3. Solve for x.
4. Put the x value into the first equation to find y.

This method will always work, although in some cases it may be easier to eliminate x and solve for y. As with simple equations, x and y can have any value. They can be positive or negative, whole numbers or fractions.

Problems

Equations are used for solving problems. Pairs of equations are used when there is a pair of unknowns.

Example: A fairground employs men and boys during the summer. One day's wages for 3 men and 2 boys are £14. For 1 man and 5 boys the wages are £9. Find the daily wage for men and boys.

Let the men earn £x each day.
Let the boys earn £y each day.

$$3x + 2y = 14 \quad \ldots\ldots\ldots\ldots (1)$$
$$x + 5y = 9 \quad \ldots\ldots\ldots\ldots (2)$$

Multiply (1) by 5 $\quad 15x + 10y = 70 \quad \ldots\ldots\ldots\ldots (3)$

Multiply (2) by 2 $\quad 2x + 10y = 18 \quad \ldots\ldots\ldots\ldots (4)$

Subtract (4) from (3) $\quad 13x = 52$

$$x = 4.$$

Substitute $x = 4$ in (1) $\quad 3 \times 4 + 2y = 14$

$$12 + 2y = 14$$
$$2y = 2$$
$$y = 1.$$

Check in (2)
LHS $4 + 5(1) = 4 + 5 = 9$. RHS 9.

$\therefore$ The men earn £4 each day and the boys earn £1 each day.

Exercise 36a

Solve the following pairs of equations.

(1) $3x + 2y = 17$
$2x + 3y = 13$

(2) $3x + 4y = 10$
$2x - 3y = 1$

(3) $4x + 3y = 14$
$3x - 2y = 2$

(4) $3x + y = 8$
$2x - 3y = 9$

(5) $3x + 5y = 13$
$5x - 3y = -1$

(6) $4x - y = 6$
$3x + 2y = 10$

(7) $3x + 2y = 5$
$x - 5y = -4$

(8) $3x - y = 8$
$x - 2y = 1$

(9) $2x + y = 4$
$x - 2y = -3$

(10) $x + 2y = 3$
$x - 3y = 8$

(11) $x + 2y = 5$
$2x + 3y = 6$

(12) $x - y = 1$
$3x - 2y = 1$

(13) $3x + 2y = 4$
$2x - 6y = -1$

(14) $4x + y = 2$
$8x + 3y = 5$

(15) $2x - 3y = -4$
$-x + 2y = 2$

(16) $4x - 2y = 1$
$3x + 5y = 4$

(17) $x - 2y = 4$
$3x + 4y = 7$

(18) $x + 3y = 2$
$3x - 3y = -2$

(19) $2x + 2y = 1$
$x - y = -1$

(20) $9x + 3y = 4$
$3x - 6y = -1$

(21) The sum of two numbers is 10 and their difference is 6. Make up a pair of equations and solve them to find the two numbers.

(22) Marcus is one year older than Philip and their ages add up to 15. Make up a pair of equations and solve them to find their ages.

(23) A bag contains 5p and 10p pieces. There are 14 coins in all and their value is £1. Make up a pair of equations and solve them to find the number of each type of coin.

(24) A scale pan contains 1 g weights and 5 g weights. In all there are 13 weights and the total weight is 37 g. Make up a pair of equations and solve them to find the number of weights of each kind.

(25) Two books have a total of 500 pages. One book has 350 pages more than the other. Make up a pair of equations and solve them to find the number of pages in each book.

Exercise 36b

(1) (a) Find the value of $3x + 4y$ if $x = 2$, $y = \frac{3}{4}$.

(b) Solve the equations $4x + 3y = 10$
$3x - y = 1$.

(c) A man is 22 years older than his son, and their total age is 48 years. Make up a pair of equations and solve them to find the two ages.

(2) (a) Find the value of $2x - 3y$ if $x = -1$ and $y = 2$.

(b) Solve the equations $2x + 7y = 1$
$x + 5y = 2$.

(c) The sum of two numbers is 5 and their difference is 15. Make up a pair of equations and solve them to find the two numbers.

(3) (a) Find the value of $5x + 3y$ if $x = 3$ and $y = \frac{2}{3}$.

(b) Solve the equations $2x + 3y = -2$

$$4x - y = 3.$$

(c) The length of a rectangle is 2 m more than its width and the perimeter is 8 m. Make up a pair of equations and solve them to find the length and breadth of the rectangle.

(4) (a) Find the value of $4x - 2y$ if $x = 2\frac{1}{2}$ and $y = -6$.

(b) Solve the equations $2x + y = 1$

$$4x + 3y = 4.$$

(c) The sum of the number of edges and faces of a solid is 20. The difference between the number of edges and faces is 4. Make up a pair of equations and solve them to find the number of edges and faces. Could this be a cube?

(5) (a) Say where $x = 2$, $y = -3$ is the solution of the equations

$$4x - 5y = 23$$
$$3x - y = 3.$$

(b) Solve the equations $4x - 5y = 3$

$$3x - y = 5.$$

(c) A cup and saucer together cost 25p and the cup costs 5p more than the saucer. Make up a pair of equations and solve them to find the cost of the cup and the cost of the saucer.

Section G
37 Ratio and Proportion

Fractions and Ratios

In the figure below, AB = 3 cm and AC = 8 cm. If we compare the lengths of AB and AC, we say that AB and AC are in the *ratio* of 3 to 8. This is written as
AB : AC = 3 : 8.

3 cm 5 cm
A B C

We also know that AB is $\frac{3}{8}$ of AC.

$$\frac{AB}{AC} = \frac{3}{8}.$$

What fraction of AC is BC? What is the ratio BC : AC? What is the ratio of AB : BC?

Fractions and ratios are often used in problems in arithmetic, as the following example shows.

Example: A bag contains red marbles and green marbles in the ratio 3 : 5. If the total number of marbles is 24, find how many there are of each colour.

---

red marbles : green marbles = 3 : 5

$\therefore$ the fraction of red marbles is $\dfrac{3}{3+5} = \frac{3}{8}$

and the fraction of green marbles is $\dfrac{5}{3+5} = \frac{5}{8}$.

$\therefore$ the number of red marbles is $\frac{3}{8} \times 24 = 9$

and the number of green marbles is $\frac{5}{8} \times 24 = 15$.

Check: 9 : 15 = 3 : 5.

Direct Proportion

If the price of a bicycle is increased by 20%, the new price can be found by adding 20% to the old price. The calculation is often easier if it is seen that the price is being increased in the ratio 120 : 100. The new price is $\frac{120}{100}$ of the old price.

Example: The price of a £40 bicycle is increased by 20%. Find the new price.

The price is increased in the ratio 120 : 100.

$$\therefore \text{ new price is } £40 \times \frac{120}{100}$$

$$= £40 \times \frac{6}{5}$$

$$= £48.$$

Inverse Proportion

If a number of children share a bottle of milk equally between themselves, the amount that each child gets will depend on the number of children. More children, less milk each.

Example: Five children expect to share two bottles of milk equally between themselves and to get 240 ml each. If there is one extra child and the milk is shared equally, how much does each receive?

The number of children has increased in the ratio 6 : 5.

$$\therefore \text{ each child gets } 240 \times \frac{5}{6} \text{ ml}$$

$$= 200 \text{ ml.}$$

Exercise 37a

(1) Increase 100 in the ratio 5 : 4.
(2) Increase 81 in the ratio 10 : 9.

(3) Increase £42 in the ratio 9 : 7.
(4) Increase £1 in the ratio 21 : 20.
(5) Increase 5.4 m in the ratio 11 : 6.
(6) Increase 7.5 cm in the ratio 7 : 5.
(7) Increase 63 kg in the ratio 9 : 7.
(8) Increase 45 g in the ratio 12 : 5.
(9) Increase 4 weeks in the ratio 10 : 7.
(10) Increase $\frac{4}{5}$ in the ratio 5 : 4.
(11) Increase £40 by 15%.
(12) Increase £475 by 20%.
(13) Increase 500 by 23%.
(14) Increase 20p by 5%.
(15) Increase 5 litres by 150%.
(16) Decrease £100 by 13%.
(17) Decrease £2.50 by 4%.
(18) Decrease 4 m by 7%.
(19) Decrease 72 kg by 25%.
(20) Decrease £50 by 70%.
(21) A bag contains white marbles and black marbles in the ratio of 6 : 5. If there are a total of 22 marbles, find how many there are of each colour.
(22) In class 5X there are 35 pupils. The boys and girls are in the ratio of 4 : 3. Find how many boys and how many girls there are in the class.
(23) 63 candidates enter for an examination. If the ratio of passes to fails is 5 : 2, find the number of passes and fails.
(24) There are 84 men and women on a coach trip. If the numbers of men and women are in the ratio 5 : 7, find how many there are of each.
(25) In a right-angled triangle, the sizes of the smaller angles are in the ratio 3 : 2. Calculate the angles of the triangle.
(26) There are 33 maths and science text books on a shelf. If the ratio of maths books to science books is 7 : 4, find how many there are of each.
(27) In a game of chess there are 26 pieces left on the board. If the numbers of black and white pieces are in the ratio 6 : 7, find how many there are of each.
(28) If a car travels 150 km on 15 litres of petrol, how far can it travel on 17 l?
(29) If 35 foreign stamps cost 42p how much would 40 of the same stamp cost?
(30) If 8 bottles of milk cost 44p what is the cost of 10 bottles?

(31) A man earns £50 for ten days work. How much does he earn in 7 days?
(32) If 15 litres of milk fill 25 bottles, how many bottles would 27 l fill?
(33) A line of eighty 10p pieces stretches 228 cm. How far do 120 stretch?
(34) A man's heart beats 600 times in 8 minutes. How many times does it beat in 1 hour at the same rate?
(35) A journey takes $4\frac{3}{4}$ hours. How long will it take at half the speed?
(36) 100 pencils can be bought for £2.98. How many can be bought for the same amount if the price is increased by 25%?
(37) A milk churn holds enough milk to fill 90 litre bottles. How many 600 ml bottles can be filled from the same churn?
(38) A car travels 360 km on a tank full of petrol. How far will it travel if the fuel consumption is increased by one third?
(39) A man's pulse rate is 72 per minute. How long will 72 beats take if the rate is increased by $\frac{1}{11}$?
(40) A journey at constant speed takes $3\frac{3}{4}$ hours. How long will it take if the speed is increased by $12\frac{1}{2}$%?

Exercise 37b

(1) Three men go on a camping holiday and take all their food with them. If they have enough food to last for 8 days, calculate how long the food would last if
(a) they eat 20% less of everything per day
(b) they had taken 25% more
(c) there had been 4 men instead of 3.

(2) A theatre sells 612 tickets at 70p each.
(a) If the numbers of men and women in the audience are in the ratio 49 : 53, find how many there are of each.
(b) Find how many tickets could be bought at 84p for the same total cost.
(c) Find how many tickets could be bought at 72p for the same total cost.

(3) In class 2 J there are 35 boys and girls. The ratio of those staying to school lunch and those going home is 3 : 2. The total cost for their school meals is £2.52.
(a) Find the number who stay to school lunch.
(b) Find the total cost of the same lunches if the price is increased by 50%.

(c) Find how many lunches can be bought at the new price for the same total cost.

(d) Find how many people are left in the class if the total number is reduced by 20%.

(4) A man drives 770 km and the petrol costs £12.10. The distances travelled on trunk roads and on motorways are in the ratio 4:7 and the total time for the journey is $10\frac{1}{2}$ hours. Calculate

(a) the distance travelled on motorways

(b) the cost of the return journey if the price of petrol is increased by 10%

(c) the time taken for the return journey if the average speed is increased by 5%

(d) the distance which could be travelled at the new cost of petrol for the same money.

(5) A man buys a sack of Grasso lawn fertiliser for 48p, which is enough for a 50 m^2 lawn, used at a rate of 30 g/m^2.

(a) Find how much Grasso the man uses per m^2 if he spreads it evenly over his lawn which is $1\frac{1}{4}$ times as large as the recommended area.

(b) Find the cost of using Turfex at the same recommended rate if the same size sack costs 25% more.

(c) Grasso also make a larger sack containing 150% more fertiliser. Find the area of lawn that this sack full would cover at the same recommended rate of use.

38 Slide Rule

How it Works

If two scales are marked as shown they could be used for adding numbers.

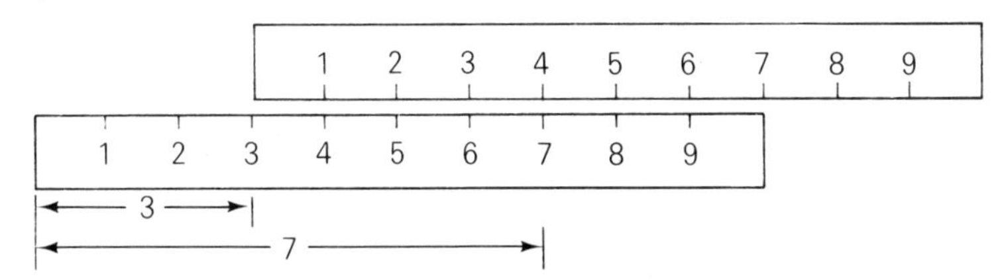

A slide rule does the same kind of thing but adds logarithms, so that it can be used for quick multiplication. Look at the scale below. It is marked in two ways. The top scale shows the line marked in tenths. The bottom scale shows the logarithms of numbers, which you can check by using your log tables.

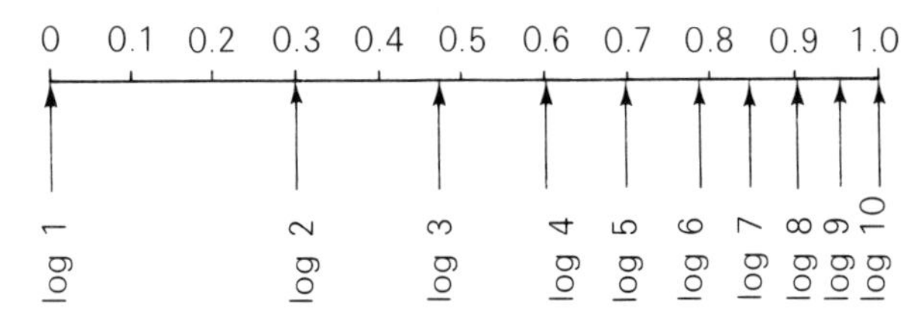

If we put two of these log scales together so as to add log 2 and log 3, we get log 6.

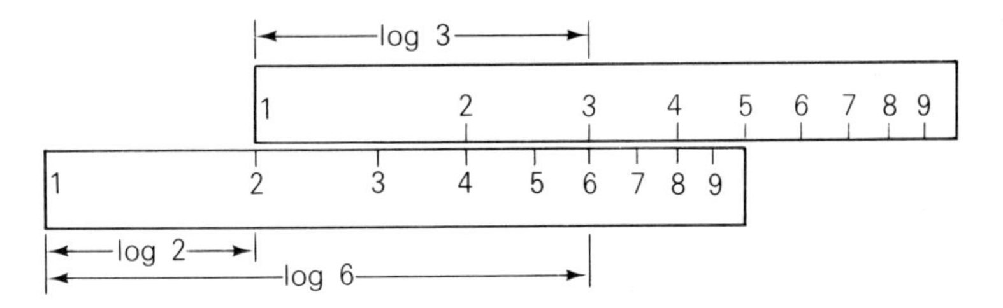

Slide rules usually have several scales. The following programme shows how to use the scales for multiplication and division. These are called the C and D scales.

Multiplication

Answers	
↓	Your slide rule is in three parts. The part shown in the figure is called the *cursor*. This is not needed yet. If you can remove the cursor, do so. If you cannot, move it to the right-hand end of the rule where it will not be in the way. Look at the left-hand end of the scales and find the C and D scales. The C scale is marked on the *slide*. Move the slide so that the 1 on the C scale is in line with the 1 on the D scale. Which number on the D scale is opposite the 4 on the C scale?
4	Which number of the D scale is opposite π on the C scale?
π	Move the slide so that the 1 on the C scale is opposite the 2 on the D scale. What number on the D scale is opposite 4 on the C scale?
8	This shows that $2 \times 4 = 8$. What number on the D scale is opposite 2.5 on the C scale?
5	This shows that $2 \times 2.5 = 5$. Be careful when reading decimals on the rule. The smallest divisions of the scale between 1 and 2 are not the same value as the smallest divisions between 4 and 5. Move the slide so that the 1 of the C scale is opposite the 3 on the D scale. What number on the D scale is opposite 2.4 on the C scale? (You may find that putting the line of the cursor over the 2.4 mark of the C scale will make it easier to read the scale.)
7.2	Use your slide rule to multiply $2.1 \times \pi$.
6.6	Calculate 1.7×2.7, correct to 1 decimal place.

4.6	Calculate 3.5×1.8, correct to 1 decimal place.
6.3	Now do Exercise 38a, questions 1–10, and then go on with the programme.

Answers ↓	
	Put the 1 of the C scale in line with the 2 on the D scale. What number on the D scale is in line with 2 on the C scale?
4	This shows that $2 \times 2 = 4$. What number on the D scale is in line with 3 on the C scale?
6	This shows that $2 \times 3 = 6$. What numbers on the D scale are in line with 4 and 5 on the C scale?
8 and 10	By keeping the slide in this position, we can multiply any number by 2 up to $5 \times 2 = 10$. The 6 on the C scale is past the end of the D scale. To get $6 \times 2 = 12$, move the slide so that the 10 of the C scale is in line with the 2 of the D scale. What number on the D scale is opposite 6 on the C scale?
1.2	This looks as though it tells us that $6 \times 2 = 1.2$. The numbers are correct, but we must ignore the decimal point. Then $6 \times 2 = 12$. In all calculations, the slide rule will give you the figures but will not give you the size of the number. You have to put the decimal point in the correct place yourself or decide how many zeros there should be at the beginning or end of the number. Move the slide so that the 10 of the C scale is in line with the 3 of the D scale. What number on the D scale is opposite 5 on the C scale?
1.5	What is the value of 3×5?
15	What number on the D scale is opposite 3.5 on the C scale?
1.05	What is the value of 3×3.5?
10.5	Move the slide so that the 10 of the C scale is opposite 3.3 on the D scale. What number on the D scale is opposite 4.3 on the C scale?
1.42	What is the value of 3.3×4.3 correct to 2 significant figures?

14	Use your slide rule to calculate 6.6×3.3 correct to 2 significant figures.
22	What is the value of 8.2×6.1 correct to 2 significant figures?
50	What is the value of 5.5×5.1 correct to 2 significant figures?
28	Now do Exercise 38a, questions 11–20, and then go on with the programme.

Answers ↓	
	To multiply 21×41, first use your slide rule to find 2.1×4.1 correct to 2 significant figures. What do you get?
8.6	21×41 is roughly $20 \times 40 = 800$. What is 21×41 correct to 2 significant figures?
860	To multiply 0.2×370, first find 2×3.7. What do you get?
7.4	0.2×370 is roughly $0.2 \times 400 = 80$. What is the value of 0.2×370 correct to 2 significant figures?
74	What is the value of 5.1×450 correct to 2 significant figures?
2 300	What is the value of 36×42 correct to 2 significant figures?
1 500	What is the value of 220×3.6 correct to 3 significant figures?
790	Now do Exercise 38a, questions 21–30.

Division

Answers ↓	
	To calculate $8 \div 4$, move the cursor so that the line of the cursor is over the 8 of the D scale. Now move the 4 on the C scale to the cursor line. What number on the D scale is in line with the end of the C scale?

2	This shows that $8 \div 4 = 2$. Notice that the slide is in the same position as for $2 \times 4 = 8$. To find $4.3 \div 5.1$ move the cursor so that the line is over 4.3 on the D scale. Now move the slide so that the 5.1 is under the cursor line. What number on the D scale is in line with the end of the C scale, correct to 2 significant figures? (Notice that you have to use the other end of the C scale this time.)
8.4	$4.3 \div 5.1$ is roughly $4 \div 5 = 0.8$. What is the value of $4.3 \div 5.1$ correct to 2 significant figures?
0.84	To find $460 \div 22$, move the cursor line to 4.6 and move the slide so that 2.2 is under the cursor line. What number on the D scale is in line with the end of the C scale, correct to 2 significant figures?
2.1	$460 \div 22$ is roughly $500 \div 20 = 25$. What is the value of $460 \div 22$ correct to 2 significant figures?
21	What is the value of $27 \div 41$ correct to 3 significant figures?
0.66	What is the value of $340 \div 5.3$ correct to 3 significant figures?
64	What is the value of $76 \div 36$ correct to 3 significant figures?
2.1.	

Most slide rules have several other scales which make calculations easier, but this will depend on the type of slide rule which you are using.

Exercise 38a

Calculate the following, correct to 2 significant figures.

(1) 3.4×2.8
(2) 1.7×2.4
(3) 2.9×2.6
(4) 5.9×1.4
(5) 2.2×2.8
(6) 1.05×6.2
(7) 7.9×1.2
(8) 3π
(9) 2.9×1.4
(10) 4.4×1.6
(11) 7.9×2.7
(12) 3.7×7.8
(13) 8.8×2.3
(14) 5.6×9.2
(15) 1.1×9.1
(16) 3.3×6.1
(17) 7.6×4.2
(18) 3.7×7.3
(19) 2.8×4.3
(20) 9.6×4.7
(21) 17×42
(22) 410×2.1
(23) 0.24×50
(24) 6.3×78

(25) 23×39
(26) 42×27
(27) 33×1.4
(28) 86×79
(29) 32×0.7
(30) 0.2×560
(31) $9.3 \div 2.6$
(32) $7.8 \div 1.7$
(33) $5.9 \div 3.4$
(34) $36 \div 4.3$
(35) $27 \div 2.9$
(36) $4.6 \div 8.2$
(37) $110 \div 31$
(38) $45 \div 76$
(39) $230 \div 51$
(40) $41 \div 7.2$
(41) $1.3 \times 2.7 \times 4.1$
(42) $3.6 \times 8.2 \times 9.7$
(43) $4.6 \times 5.3 \times 1.6$
(44) $3.3 \times 2.1 \times 4.5$
(45) $2 \times \pi \times 6.7$
(46) $3\pi \times 4.4$
(47) $\pi \times (3.2)^2$
(48) $5.7 \times 3.2 \times 25$
(49) $63 \times 21 \times 1.5$
(50) $3.7 \times 21 \times 4.5$.

Exercise 38b

Use your slide rule for the following questions and give your answers correct to 2 significant figures.

(1) If $x = 3.4$ and $y = 7.6$, calculate
  (a) xy
  (b) $x \div y$
  (c) x^2.
(2) Calculate the circumference of the circles with radii
  (a) 5.3 cm
  (b) 6.9 cm
  (c) 2.9 cm. ($C = 2\pi r$.)
(3) If $a = 5.9$, $b = 67$ and $c = 21$, calculate
  (a) abc
  (b) $c \div b$
  (c) $\dfrac{ab}{c}$
  (d) $a^2 \div c$.
(4) An examination is marked out of 80. In order to give the marks as a percentage they are each increased in the ratio 5:4. Find the increased mark corresponding to
  (a) 23
  (b) 42
  (c) 67
  (d) 71.
(5) A sphere has radius 2.4 cm. Calculate
  (a) the circumference ($C = 2\pi r$)
  (b) the surface area ($A = 4\pi r^2$)
  (c) the volume ($V = \dfrac{4\pi r^3}{3}$).

39 Quadrilaterals

No Parallel Sides

A polygon with four sides is called a *quadrilateral*. It may have parallel sides or it may not. Let us think of a quadrilateral with no parallel sides, no equal sides and no equal angles. This is an irregular quadrilateral.

There are some facts about an irregular quadrilateral which we already know. First of all, we know that it will tessellate.

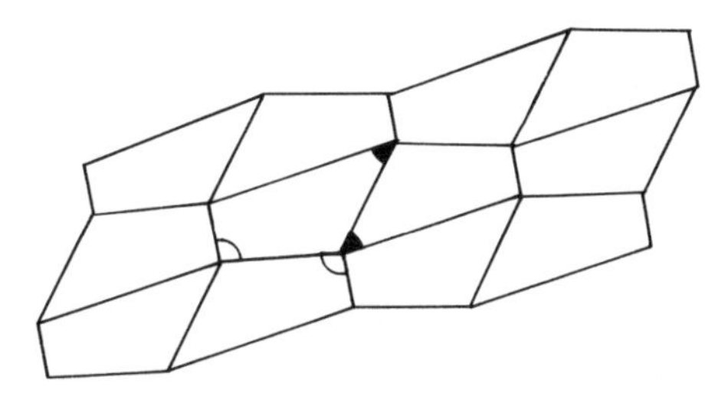

The tessellating pattern shows us that the angles of a quadrilateral add up to 360°. This is true for all quadrilaterals. Does this agree with the formula on page 122?

Kite

There is one special quadrilateral which has no sides parallel. It is the *kite*.

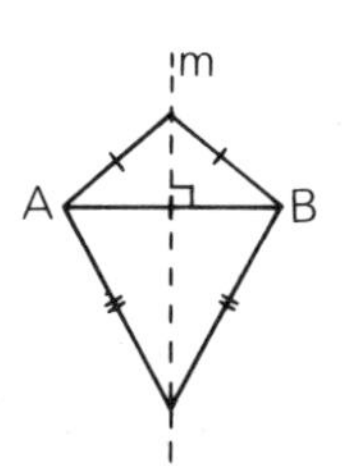

A kite has two pairs of adjacent sides equal. One of its diagonals is a line of symmetry. In the figure, B is the image of A in the mirror line. Therefore, AB is perpendicular to the mirror line. Also, $\angle A = \angle B$.

One Pair of Parallel Sides

A quadrilateral with one pair of parallel sides is called a *trapezium*. It is like a triangle with the top cut off. If it is an isosceles triangle, then the trapezium is an *isosceles trapezium*.

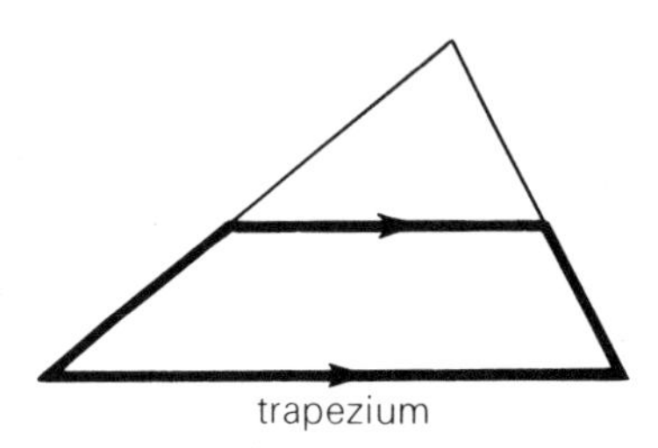

trapezium

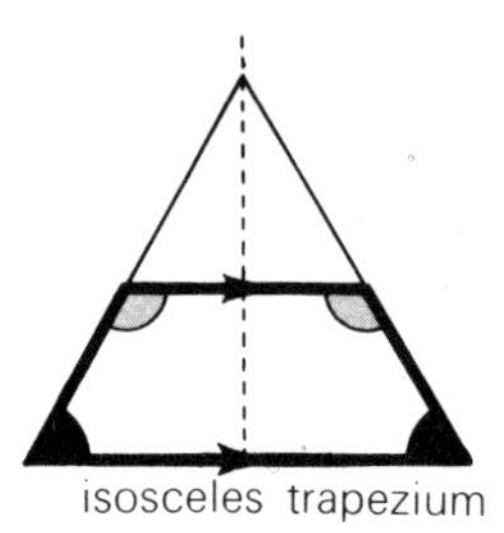

isosceles trapezium

An isosceles trapezium has two pairs of equal angles, as shown and is symmetrical.

Two Pairs of Parallel Sides

A quadrilateral with both pairs of opposite sides parallel is called a *parallelogram*.

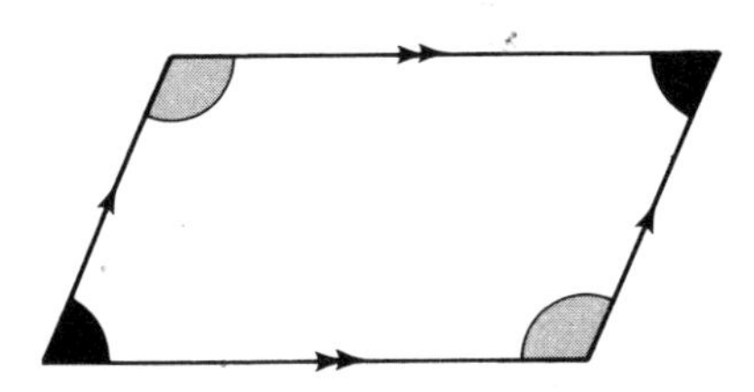

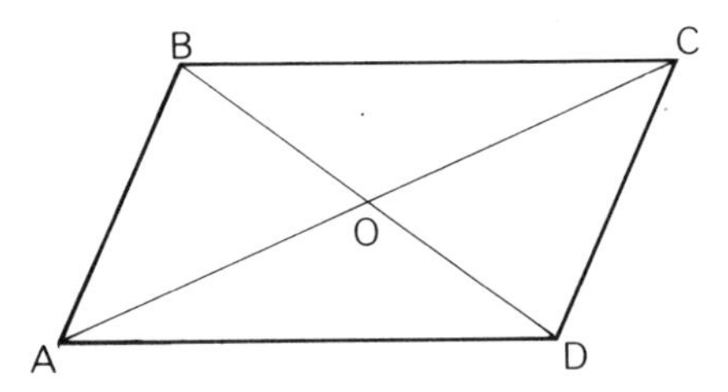

You should try to remember the following facts about parallelograms:

opposite sides are parallel and equal
opposite angles are equal
each diagonal cuts the parallelogram into halves
the two diagonals cut each other in half.

Also, notice that the two diagonals cut the parallelogram into 2 pairs of congruent triangles.

Special Parallelograms

The angles of a quadrilateral add up to 360°. If the four angles are equal, then they must each be 90°. A parallelogram with all 90° angles is called a *rectangle*. A parallelogram with all sides equal is like a diamond and is called a *rhombus*. A parallelogram which has all sides equal *and* all angles equal is called a *square*.

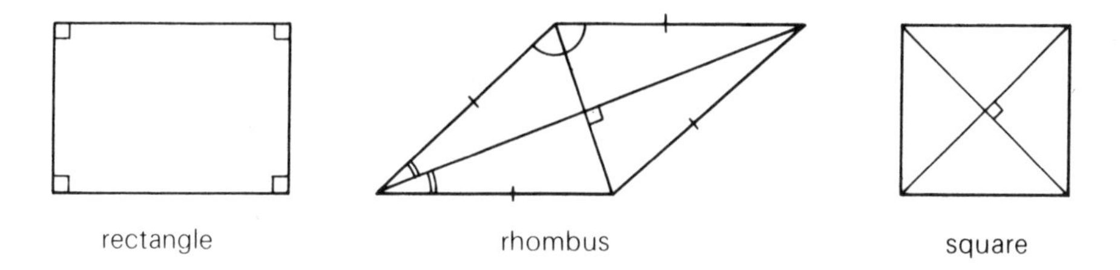

The diagonals of a rhombus are perpendicular to each other. They bisect the angles at the corners. Notice that because a square has all angles equal it is a rectangle. Because it has all sides equal, it is also a rhombus.

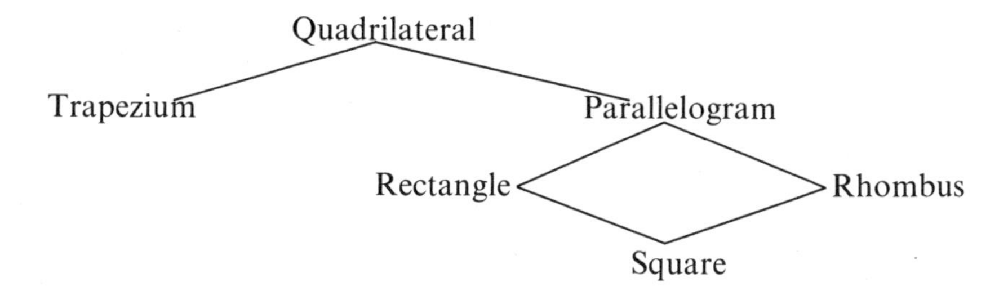

Exercise 39a

Say whether the following are true or false.

(1) The diagonals of a parallelogram are equal.
(2) The diagonals of a rectangle are equal.
(3) The diagonals of a rhombus are equal.
(4) The diagonals of an isosceles trapezium are equal.
(5) The diagonals of a rhombus are perpendicular.
(6) The diagonals of a rectangle are perpendicular.
(7) The diagonals of a square are perpendicular.
(8) The diagonals of a rectangle bisect the angles at the corners.
(9) The diagonals of a parallelogram bisect the angles at the corners.
(10) The diagonals of a square bisect the angles at the corners.
(11) Opposite angles of a parallelogram are equal.

(12) Opposite angles of a trapezium are equal.
(13) Opposite sides of a trapezium are equal.
(14) An isosceles trapezium has two sides equal.
(15) A rhombus has four equal sides.

If $\mathscr{E}$ = {quadrilaterals}, P = {parallelograms}, R = {rectangles}, D = {rhombuses}, S = {squares}, say whether the following statements are true or false:

(16) $S \subset R$ (17) $S \subset D$ (18) $D \subset R$ (19) $R \not\subset \mathscr{E}$ (20) $P \subset S$.

(21) In figure (a), calculate (i) $\angle x$, (ii) $\angle y$, (iii) $\angle z$.

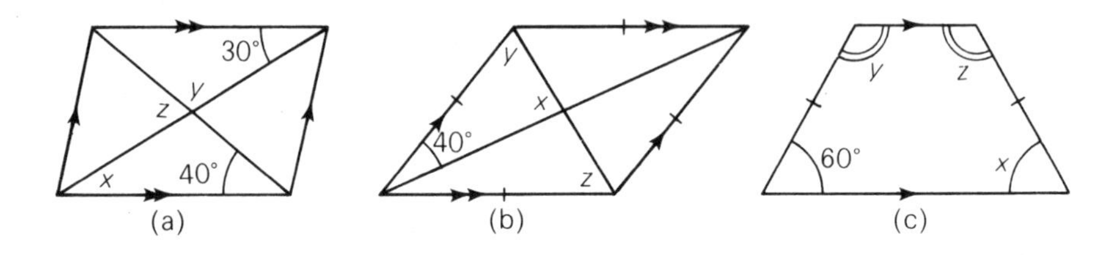

(a) (b) (c)

(22) In figure (b), calculate (i) $\angle x$, (ii) $\angle y$, (iii) $\angle z$.
(23) In figure (c), calculate (i) $\angle x$, (ii) $\angle y$, (iii) $\angle z$.
(24) Draw a triangle ABC with AB = 7 cm, BC = 7 cm and AC = 5 cm. Mark a point P on AB so that AP = 3 cm. Mark a point Q on BC so that QC = 3 cm. Join PQ. What kind of figure is APQC?
(25) Draw a quadrilateral ABCD in which AB = 4 cm, BC = 4 cm, AC = 3 cm, CD = 7 cm, DA = 7 cm. What kind of figure is this?
(26) Draw a parallelogram PQRS in which PQ = 4 cm, PS = 6 cm and $\angle$P = 60°. Measure QS correct to the nearest mm.
(27) Draw a rhombus with diagonals 6 cm and 4 cm. Measure the length of the side of the rhombus correct to the nearest mm.
(28) Draw a square with sides 5 cm long. Measure the length of the diagonal correct to the nearest mm.
(29) Draw a square with diagonals 5 cm long. Measure the length of the sides of the square correct to the nearest mm.
(30) Draw a pattern of tessellating kites. Can you mark a line of symmetry on your pattern?

Exercise 39b

(1) (a) A quadrilateral has both pairs of opposite sides equal and parallel, its diagonals are equal and bisect each other, and all its angles are 90°. What kind of quadrilateral is it?

(b) One angle of a rhombus is 45°. Calculate the other angles.

(c) Draw a rhombus ABCD with AB = 5 cm, BD = 6 cm. Measure AC correct to the nearest mm.

(d) Mark the lines of symmetry on your rhombus.

(2) (a) Say whether the following statements are true or false: (i) a square is a rhombus (ii) a rectangle is a square (iii) a trapezium is a parallelogram.

(b) Draw a square of side 3 cm and mark all the axes of symmetry.

(c) $\triangle$ABC has AB = BC = 3 cm and AC = 4 cm. $\triangle$ACD has AD = CD = 5 cm and AC = 4 cm. A quadrilateral ABCD is formed by putting these two triangles together. Draw the quadrilateral ABCD and say what kind of quadrilateral it is.

(3) If $\mathscr{E}$ = {quadrilaterals}, P = {parallelograms}, D = {rhombuses}, R = {rectangles}, S = {squares}, T = {trapeziums},

(a) Draw a Venn diagram showing $\mathscr{E}$, P and T.

(b) Draw a Venn diagram showing P, D, R and S.

(c) Say whether the following statements are true or false:
(i) $T \cap R \neq \phi$ (ii) $P \cup T = \mathscr{E}$ (iii) $R \cap D = S$.

(d) The vertices of a parallelogram lie on a circle. What kind of parallelogram is it?

(4) (a) What is the name given to a parallelogram with all sides equal and all angles equal?

(b) What kind of parallelogram has its diagonals at right angles?

(c) Two angles of a kite are 40° and 100°. What size are the other angles?

(d) The diagonals of a kite ABCD intersect at X. If AX = XC = 4 cm, BX = 3 cm and XD = 7 cm, draw the kite and calculate its area.

(5) (a) Draw a Venn diagram showing {rectangles} and {rhombuses}. Say what {rectangles} $\cap$ {rhombuses} represents.

(b) Draw a Venn diagram showing T = {trapeziums}, I = {isosceles trapeziums} and K = {kites}. Say whether the following statements are true or false:
(i) $I \cap T = I$ (ii) $T \cap K = I$ (iii) $I \cup T = T$.

(c) ABCD is a trapezium in which AD $\parallel$ BC, AD = 7 cm and BC = 3 cm. BC and AD are 4 cm apart. Calculate the area of (i) $\triangle$ABD (ii) $\triangle$BCD (iii) the trapezium ABCD.

(d) Draw the trapezium in (c) if $\triangle$ABD is isosceles.

40 Factors

Common Factors

If we multiply brackets, each term in the brackets is multiplied by the number outside (see chapter 12). For example, $3(x + y) = 3x + 3y$. The factors of $3x$ are 3 and x. The factors of $3y$ are 3 and y. Both terms have a factor of 3. This is a *common factor* because it is a factor of both terms. Going back the other way, we can say that $3x + 3y = 3(x + y)$. In the same way

$$ab + ac = a(b + c)$$
$$2xy - 6x^2 = 2x(y - 3x).$$

Factors in Pairs

To find the factors of an expression with four terms, look for a pair of terms with a common factor.

Example: Find the factors of $xy - 3x + 2y - 6$.

1. The first two terms have a common factor of x.

$xy - 3x + 2y - 6$
$= (xy - 3x) + 2y - 6$

2. Take the x outside the brackets.

$= x(y - 3) + 2y - 6$

3. $2y$ and 6 have a common factor of 2. Take the 2 outside the bracket.

$= x(y - 3) + 2(y - 3)$

4. $x(y - 3)$ and $2(y - 3)$ have a common factor of $(y - 3)$. Take out $(y - 3)$.

$= (y - 3)(x + 2).$

Quadratics

If two brackets such as $(5x + 2)$ and $(x + 3)$ are multiplied together, the result is a *quadratic expression.*

$$(5x + 2)(x + 3) = 5x^2 + 15x + 2x + 6$$
$$= 5x^2 + 17x + 6.$$

To factorise $5x^2 + 17x + 6$, we must work back the other way. First look at the coefficient of x^2 which is 5, and the constant, which is 6.

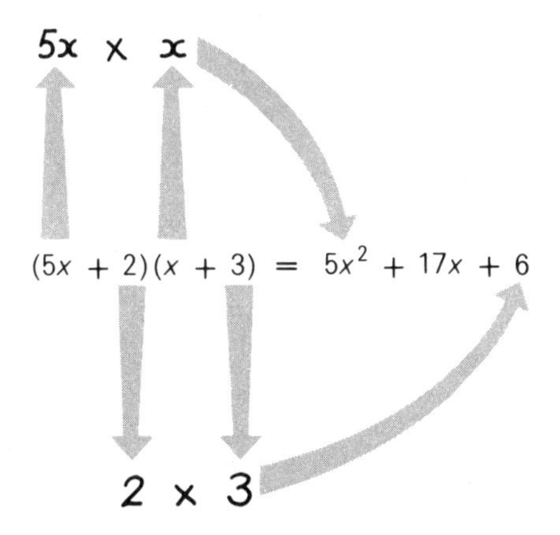

Because the factors of $5x^2$ are $5x$ and x, we know that the first term in one bracket must be $5x$ and that the first term in the other bracket must be x.

$(5x \blacksquare)(x \blacksquare) = 5x^2 \blacksquare$.

Because the factors of 6 are 1 and 6 or 2 and 3 we know that the second numbers are either 1 and 6 or 2 and 3. To find out which pair of factors to use, we must try them to see which pair works.

Try 1 and 6 $(5x + 1)(x + 6) = 5x^2 + 31x + 6.$
Try 6 and 1 $(5x + 6)(x + 1) = 5x^2 + 11x + 7.$
Try 2 and 3 $(5x + 2)(x + 3) = 5x^2 + 17x + 6.$

We then see that the factors of $5x^2 + 17x + 6$ are $(5x + 2)$ and $(x + 3)$. We knew that the signs must be the same in both brackets because the constant is $+6$. If the constant is positive, the signs in the brackets are the same. If the constant is negative, the signs in the brackets are different.

Example: Factorise $2x^2 - 5x - 3$.

1. The factors of $2x^2$ are $2x$ and x.

$2x^2 - 5x - 3 = (2x \blacksquare)(x \blacksquare)$

2. The factors of -3 are $\left\{\begin{matrix} -1 \text{ and } 3 \\ 1 \text{ and } -3 \end{matrix}\right\}$

rough work

$(2x-1)(x+3) = 2x^2+2x-3$

$(2x+1)(x-3) = 2x^2-5x-3$ ✓

Try pairs of factors to get the middle term right.

3. The factors are $(2x + 1)$ and $(x - 3)$.

$2x^2 - 5x - 3 = (2x + 1)(x - 3)$.

Perfect Squares

If you multiply two brackets which are the same, the result is a *perfect square*.

$$(a + b)^2 = (a + b)(a + b) = a^2 + 2ab + b^2$$
$$(a - b)^2 = (a - b)(a - b) = a^2 - 2ab + b^2.$$

Difference of Two Squares

If you multiply two brackets which are the same except for the sign, the product only has two terms. It is the difference of two squares.

$$\begin{aligned}(a + b)(a - b) &= a^2 - ab + ba - b^2 \\ &= a^2 - b^2.\end{aligned}$$

This means that an expression which is the difference of two squares can easily be factorised.

Example: Factorise $9x^2 - 16y^2$.

1. $9x^2$ is $(3x)^2$ so the first term in each bracket is $3x$.

$9x^2 - 16y^2 = (3x + \blacksquare)(3x - \blacksquare)$

2. $16y^2$ is $(4y)^2$ so the second term in each bracket is $4y$.

$9x^2 - 16y^2 = (3x + 4y)(3x - 4y)$.

Exercise 40a

Factorise

(1) $ax + ay$
(2) $3x + 3z$
(3) $5a + 10b$
(4) $3x - 6y$
(5) $2b - 2ab$
(6) $6x + 10y$
(7) $21xy - 6x^2$
(8) $6x - 2$
(9) $4x + 14x^2$
(10) $15x - 5$
(11) $pr + ps + qr + qs$
(12) $ac + ad + bc + bd$
(13) $wy + wz + xy + xz$
(14) $ab - ad - cb + cd$
(15) $2x^2 + 2xy + x + y$
(16) $x + y + x^2 + xy$
(17) $2ac - 2ad + bc - bd$
(18) $ac + 2ad + 2bc + 4bd$
(19) $2ax - x + 6ay - 3y$
(20) $3xy - 2x + 3y - 2$
(21) $x^2 + 3x + 2$
(22) $x^2 + 5x + 6$
(23) $x^2 - x - 6$
(24) $x^2 - 2x - 3$
(25) $x^2 + x - 2$
(26) $x^2 - 6x - 7$
(27) $x^2 + 10x + 25$
(28) $x^2 + 6x + 9$
(29) $x^2 - 2x + 1$
(30) $x^2 - 4x + 4$
(31) $3x^2 + 4x + 1$
(32) $5x^2 + 6x + 1$
(33) $2x^2 - 5x - 3$
(34) $2x^2 - 3x + 1$
(35) $2x^2 + 5x + 3$
(36) $3x^2 + 5x - 2$
(37) $2x^2 + 7x + 3$
(38) $3x^2 + 8x + 5$
(39) $2x^2 + 15x + 7$
(40) $2x^2 - 5x + 3$
(41) $2x^2 + 13x + 6$
(42) $2x^2 + 7x + 6$
(43) $3x^2 + 4x - 4$
(44) $2x^2 - x - 6$
(45) $3x^2 + 13x + 4$
(46) $2x^2 - 11x - 6$
(47) $3x^2 + 17x + 10$
(48) $3x^2 + x - 10$
(49) $3x^2 - 4x - 4$
(50) $3x^2 + 13x + 4$
(51) $a^2 - b^2$
(52) $x^2 - y^2$
(53) $x^2 - 4y^2$
(54) $a^2 - 9b^2$
(55) $4x^2 - 9y^2$
(56) $9a^2 - 4b^2$
(57) $16p^2 - 9q^2$
(58) $4s^2 - 25t^2$
(59) $25x^2 - 4y^2$
(60) $49y^2 - 64x^2$

Exercise 40b

(1) Factorise
(a) $6x^2 + 15xy$
(b) $2pr - 2ps + qr - qs$
(c) $6x^2 - 13x + 6$
(d) $9x^2 - 49y^2$.

(2) Factorise
(a) $21pq - 14pa$
(b) $xy + 2x + 3y + 6$
(c) $2x^2 - 11x - 6$
(d) $25a^2 - 81b^2$.

(3) (a) A rectangle is of length $(x + 2)$ cm and width $(x - 1)$ cm. Write an expression for the area in descending powers of x.
(b) Factorise $6xy - 4x + 3y - 2$.
(c) Factorise $3x^2 + 22x + 35$.
(d) Factorise $16x^2 - 25y^2$.

(4) Factorise
(a) $8ap + 2aq$
(b) $x^2 - xz + xy - yz$
(c) $2x^2 - 7x + 5$
(d) $81a^2 - b^2$.

(5) (a) A rectangle has an area of $(3x^2 + 6x)$ cm^2. If its length is $3x$ cm find its width.
(b) Factorise $3x^2 + 5x - 22$.
(c) If the area of a square is $(a^2 + 2ab + b^2)$ cm^2 what is the length of the side of the square?
(d) Factorise $p^2 - 25q^2$.

41 Commerce

Profit and Loss

Goods are made by a *manufacturer*. They are sold to the public by the shopkeeper, or *retailer*. Between the manufacturer and the retailer is the *wholesaler*. The wholesaler buys goods from the manufacturer. The retailer buys them from the wholesaler.

Some traders buy and sell goods which have already been used. These goods are called *second-hand* goods. Cars are a good example of this. The price which a dealer pays for a car is his *cost price* (CP). The amount he sells it for is the *selling price* (SP). The difference between the cost price and the selling price is the *gain* or *profit*.

Sometimes goods have to be sold for less than the cost price. In this case, instead of a profit, there is a *loss*. The profit or loss is often stated as a percentage of the cost price.

$$\text{Percentage profit} = \frac{\text{profit}}{\text{cost price}} \times 100 = \frac{\text{selling price} - \text{cost price}}{\text{cost price}} \times 100.$$

$$\text{Percentage loss} = \frac{\text{loss}}{\text{cost price}} \times 100 = \frac{\text{cost price} - \text{selling price}}{\text{cost price}} \times 100.$$

Example: A dealer buys a car for £250 and sells it for £350. Find the percentage profit.

Percentage profit is $\frac{SP - CP}{CP} \times 100$

$$= \frac{£350 - £250}{£250} \times 100 = \frac{100}{250} \times 100 = 40\%.$$

Commission and Discount

A man who sells goods to the public is a *salesman*. His wages may depend on how much he sells. If he is paid £5 for every £100 worth of goods sold, he is paid 5% *commission*.

Sometimes a retailer holds a *sale*. Goods are sold at less than the usual price. If the retailer reduces the price of the goods by 5%, then the goods are sold at 5% *discount*. Rates and other bills can sometimes be settled at a discount for prompt payment. This means that the householder pays $2\frac{1}{2}$% or 5% less than the amount stated on the bill if it is paid within 2 or 3 weeks.

Example: A £40 coat is marked at 10% discount in a sale. Find the sale price.

The coat is sold for 90% of £40.

$\therefore$ Sale price is £40 $\times \frac{9}{10} =$ £36.

Interest

If you hire a car for a week to go on holiday, you pay the car hire firm for lending you the car. If you borrow money for 2 or 3 years to buy a car, you pay the bank or finance company for lending you the money. The amount you pay for borrowing money depends on how much you borrow and how long you borrow it for. It is called *interest*. If you borrow money for one year, the interest is a fixed percentage of the amount borrowed. This percentage is called the *rate* and the amount borrowed is called the *principal*.

If you save money in a bank or with a building society, your money will be lent to people who pay interest and you in your turn will be paid interest for lending the money. If you save with a bank the interest will be about $6\frac{1}{4}$%. If you save with a building society it will be about 7%.

Simple Interest

Suppose you borrow £100 and pay interest at the rate of 7% per annum, (per annum means each year). This means that in the first year, the interest is 7% of £100, which is £7. If you continue to pay £7 interest each year then you pay *simple interest*.

If you borrow £350 for 3 years at 5% per annum, the interest you pay each year is $£350 \times \frac{5}{100}$. For 3 years the total interest paid is

$$£350 \times \frac{5}{100} \times 3 = £\frac{350 \times 5 \times 3}{100}.$$

If the principal is $£P$, the rate is $R\%$ per annum and the time is T years, then

$$\text{simple interest, } I = £\frac{PRT}{100}.$$

Example: Calculate the simple interest on £320 at 6% for 4 years.

$$\text{Simple interest is } \frac{PRT}{100} = £\frac{320 \times 6 \times 4}{100}$$

$$= £76.80.$$

Compound Interest

The principal and the interest added together give the *amount*. If the amount at the end of the one year is taken to be the principal for the next year, you pay *compound interest*.

Example: Calculate the amount after paying compound interest for 2 years at £320 at $2\frac{1}{2}\%$.

First year $P = £320$

$$I = £320 \times \frac{2\frac{1}{2}}{100} = £320 \times \frac{1}{40} = £8.$$

Amount $= P + I = £328.$

Second year $P = £328$

$$I = £328 \times \frac{2\frac{1}{2}}{100} = £\frac{328}{40} = £8.20.$$

Amount $= P + I = £336.20.$

Remember that $2\frac{1}{2}\% = \frac{1}{40}$, $7\frac{1}{2}\% = \frac{3}{40}$ and $12\frac{1}{2}\% = \frac{1}{8}$.

Exercise 41a

Calculate the percentage profit or loss in each of questions 1–10.

(1) CP £200, SP £230
(2) CP £450, SP £495
(3) CP £2.75, SP £3.30
(4) CP £5, SP £4.95
(5) CP £7.50, SP £7.80
(6) CP 80p, SP £1
(7) CP £2.50, SP £1.95
(8) CP £40, SP £39
(9) CP £16, SP £14
(10) CP £72, SP £99.

Calculate the following:

(11) 5% commission on £540
(12) $2\frac{1}{2}$% commission on £80
(13) $7\frac{1}{2}$% commission on £120
(14) $12\frac{1}{2}$% commission on £64
(15) 7% commission on £730
(16) 3% discount on £560
(17) 4% discount on £23
(18) 20% discount on £47
(19) $7\frac{1}{2}$% discount on £60
(20) 10% discount on £68.

In questions 21–30, calculate the simple interest as stated.

(21) £460 for 3 years at 5%
(22) £235 for 5 years at 4%
(23) £620 for 4 years at 3%
(24) £125 for 3 years at 7%
(25) £550 for 5 years at $2\frac{1}{2}$%
(26) £2 000 for 20 years at 11%
(27) £243 for 3 years at 6%
(28) £46 for 3 years at 10%
(29) £83 for 3 years at 9%
(30) £35 for 5 years at 11%.

In questions 31–40, calculate the amount if compound interest is paid for 2 years as stated.

(31) £100 at 10%
(32) £200 at 5%
(33) £240 at $2\frac{1}{2}$%
(34) £300 at 4%
(35) £1 000 at 7%
(36) £80 at 10%
(37) £600 at 3%
(38) £250 at 10%
(39) £450 at 20%
(40) £750 at 4%.

Exercise 41b

(1) A retailer buys a washing machine for £50 and tries to sell it for £58.
(a) What percentage profit does he expect?
(b) If it is sold in a sale at a discount of 10% on the £58, what is the selling price?
(c) Calculate the salesman's commission which is 5% of the discount selling price.

(2) A dealer buys a car for £500 and sells it for £600.
(a) Calculate the percentage profit.

(b) Calculate the salesman's commission if it is 7% of the selling price.

(c) Find the commission as a percentage of the cost price.

(3) A salesman sells goods to the value of £2 750 and earns commission at 5%.

(a) Calculate his commission.

(b) If he invests the commission for 10 years at 8% simple interest, calculate the interest paid.

(c) Calculate the amount after 2 years.

(d) Calculate the amount after compound interest is paid on £137.50 for 2 years at 8%.

(4) (a) Calculate the simple interest on £760 for 5 years at $7\frac{1}{2}$%.

(b) Calculate the compound interest on £800 for 2 years at $2\frac{1}{2}$%.

(c) If £1 000 is invested for 2 years at 5%, find the difference between the simple interest and compound interest in this time.

(5) Calculate

(a) the simple interest on £475 for 7 years at 2%

(b) 5% of £600

(c) for how long is £600 invested at 5% simple interest, if the interest is £120?

(d) the compound interest on £600 at 5% for 2 years.

42 Transformation of Formulae

Simple Interest

In chapter 41, we used the formula

$$I = \frac{PRT}{100}$$

for calculating simple interest. Because the formula tells us how to find I, we say that I is the *subject* of the formula. If we know I, P and R, but not T, we could use the same formula to find T or we could re-arrange the formula to make T the subject.

$$I = \frac{PRT}{100}.$$

Multiply both sides by 100

$$100\,I = PRT.$$

Divide both sides by PR

$$\frac{100\,I}{PR} = T$$

so that

$$T = \frac{100\,I}{PR}.$$

In the same way we could make P the subject, or R.

Changing the Subject

We can change the subject of a formula providing we remember how to solve equations. Whatever is done to one side of the formula must also be done to the other side.

Example: Make b the subject of the formula $A = \frac{1}{2}(a + b)h$.

1. Write the formula as given.
2. Multiply both sides by 2.
3. Divide both sides by h.
4. Subtract a from both sides.

$$A = \tfrac{1}{2}(a + b)h$$

$$\therefore 2A = (a + b)h$$

$$\therefore \frac{2A}{h} = a + b$$

$$\frac{2A}{h} - a = b$$

or $$b = \frac{2A}{h} - a.$$

Formulae with Squares

We can find the area of a circle if the radius is known, by using the formula $A = \pi r^2$. Can we find r if A is known? To do this we need r as the subject of the formula.

$\pi r^2 = A.$

Divide both sides by π

$$r^2 = \frac{A}{\pi}.$$

To get r from r^2, we must take the square root. If we take the square root of the left-hand side, we must also take the square root of the right-hand side, so that

$$r = \sqrt{\frac{A}{\pi}}\ .$$

Example: If $P = 3x^2y$ make x the subject of the formula.

1. Write the formula with x^2 on the LHS.
2. Divide both sides by $3y$.
3. Take the square root of both sides.

$$3x^2y = P$$

$$x^2 = \frac{P}{3y}$$

$$x = \sqrt{\frac{P}{3y}}.$$

Formulae with Square Roots

If you start with a formula which has a square root, you must square both sides.

Example: If $T = 2\pi\sqrt{\frac{l}{g}}$, make l the subject of the formula.

1. Write the formula with l on the LHS.
$$2\pi\sqrt{\frac{l}{g}} = T$$

2. Divide both sides by 2π.
$$\sqrt{\frac{l}{g}} = \frac{T}{2\pi}$$

3. Square both sides.
$$\frac{l}{g} = \frac{T^2}{4\pi^2}$$

4. Multiply both sides by g.
$$l = \frac{gT^2}{4\pi^2}.$$

Exercise 42a

Change the following formulae so that the letter in square brackets becomes the subject.

(1) $A = lb$, $[b]$
(2) $V = lhb$, $[l]$
(3) $A = \pi ab$, $[a]$
(4) $A = 2\pi rh$, $[r]$
(5) $A = \frac{1}{2}bh$, $[h]$
(6) $v = u + at$, $[t]$
(7) $P = 2l + 2b$, $[l]$
(8) $s = \frac{1}{2}(a + b + c)$, $[c]$
(9) $P = 3x + 2y$, $[x]$
(10) $P = \frac{x}{4y + 1}$, $[x]$
(11) $Q = \frac{a + b}{a - c}$, $[b]$
(12) $X = \frac{2p + q}{q}$, $[p]$
(13) $K = \frac{1}{2}mv^2$, $[m]$
(14) $E = mc^2$, $[c]$
(15) $c^2 = a^2 + b^2$, $[a]$
(16) $s = \frac{1}{2}at^2$, $[t]$
(17) $A = 4\pi r^2$, $[r]$
(18) $d = \sqrt{(2A)}$, $[A]$
(19) $v = \sqrt{(u^2 + 2as)}$, $[s]$
(20) $r = \sqrt{\frac{V}{\pi h}}$, $[V]$.

Exercise 42b

(1) If $r^2 = x^2 + y^2$.
(a) Calculate r when $x = 4$ and $y = 5$.

(b) Find an expression for x^2.
(c) Factorise your expression for x^2.
(d) Find the value of x if $r = 14$ and $y = 7$.

(2) The curved surface area of a cylinder is given by $A = 2\pi rh$.
(a) Use logarithms to calculate A if $r = 3.4$ cm and $h = 2.7$ cm (take $\pi = 3.142$).
(b) Express h in terms of A, r and constants.
(c) Calculate the value of h if $A = 75$ cm^2 and $r = 5$ cm.

(3) The surface area of a sphere is given by $S = 4\pi r^2$.
(a) Calculate S if $r = 5$ cm.
(b) Make r the subject of the formula.
(c) Calculate r if $S = 50$ cm^2.

(4) If $A = \frac{P}{100}(100 + RT)$.
(a) Calculate A if $P = 500$, $R = 4$ and $T = 5$.
(b) Make P the subject of the formula.
(c) Calculate P if $A = 975$, $R = 5$, and $T = 6$.

(5) If $Q = \frac{x^2}{y}$
(a) Calculate Q when $x = -14$ and $y = 49$.
(b) Make y the subject of the formula.
(c) Make x the subject of the formula.
(d) Calculate x when $Q = 7.6$ and $y = 3.2$.

43 Enlargement, Shear

Enlargement

When a shape is enlarged, all the angles stay the same, and ratios of distances stay the same. Suppose the kite ABCD is enlarged to A′B′C′D′.

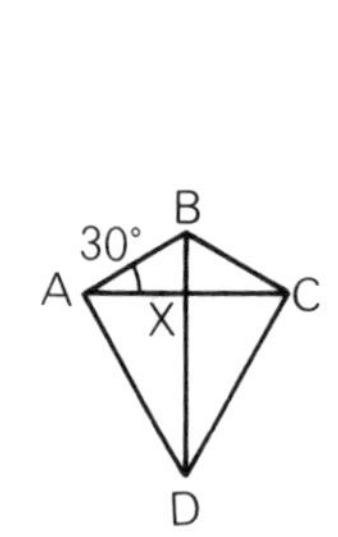

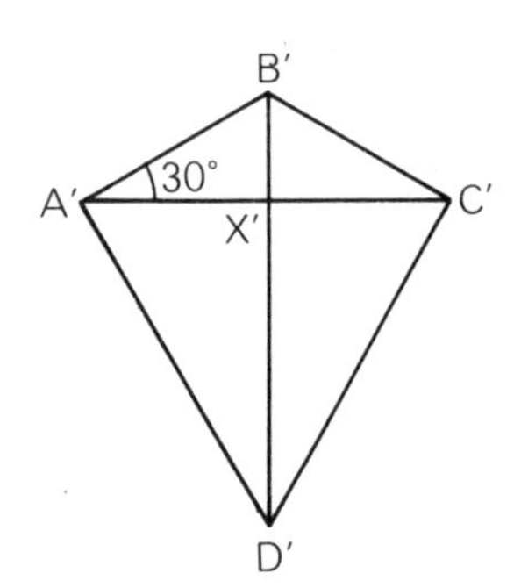

If you measure ∠BAC and ∠B′A′C′ you will find that they are both 30°. In the same way ∠BDC = ∠B′D′C′. Every angle in ABCD is equal to the corresponding angle in A′B′C′D′.

All lengths in A′B′C′D′ are twice the corresponding lengths in ABCD.

For example, suppose

BX = 1 cm, then B′X′ = 2 cm
AB = 2 cm, then A′B′ = 4 cm.

Although the distances have doubled, the ratio of the distances is the same.

$$\frac{BX}{AB} = \frac{1}{2}, \qquad \frac{B'X'}{A'B'} = \frac{1}{2}.$$

The two kites are the same shape and are said to be *similar*.

Because the lengths in A′B′C′D′ are twice the lengths in ABCD the area is $2^2 = 4$ times as large.

Solids

When solids are enlarged, angles stay the same and so do ratios of lengths as for plane figures.

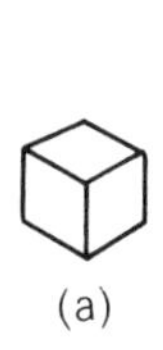
(a)

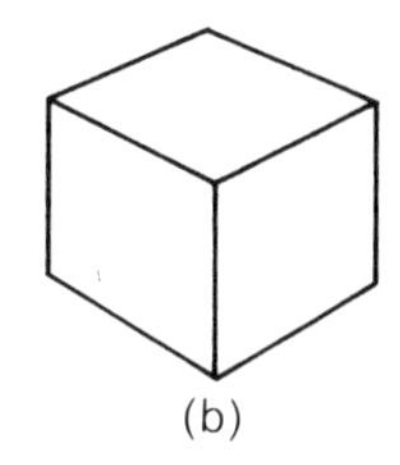
(b)

Look at the cubes (a) and (b). The length of a side of (b) is twice the length of a side of (a). The area of the surface of (b) is $2^2 = 4$ times the surface of (a). The volume of (b) is $2^3 = 8$ times the volume of (a).

The ratio of corresponding lengths of (b) and (a) is $2:1$.
The ratio of corresponding areas of (b) and (a) is $4:1$.
The ratio of corresponding volumes of (b) and (a) is $8:1$.

If a solid is enlarged so that lengths become n times as large,

the ratio of corresponding lengths is $n:1$
the ratio of corresponding areas is $n^2:1$
the ratio of corresponding volumes is $n^3:1$.

n is called the *magnification*.

Shear

Imagine a pile of postcards which is 'pushed over' as shown.

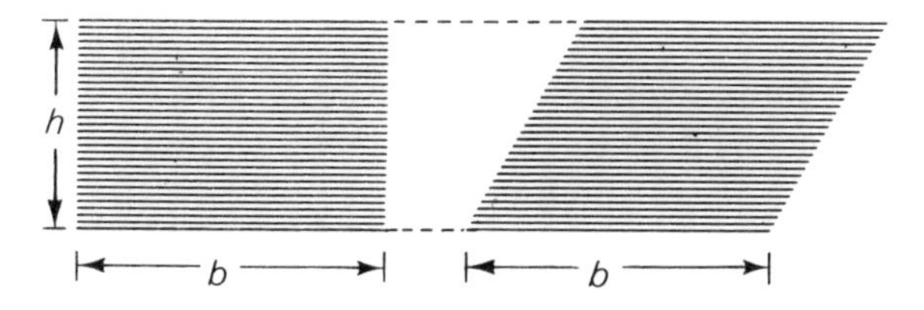

This kind of change is called a *shear*.

The end view of the pile has changed from a rectangle to a parallelogram. The amount of card showing at the end is still the same, so that the area of the parallelogram must be the same as the area of the rectangle. The area of the rectangle is $A = bh$, so the area of the parallelogram must be $A = bh$. b is the length of the base and h is the distance between the base and the side which is parallel to it.

When a shape is sheared, parallel lines stay parallel and areas are unchanged.

Exercise 43a

(1) A rectangle is enlarged so that it becomes twice as long and twice as wide. If the diagonal of the rectangle was 7 cm, what is the length of the diagonal of the enlarged rectangle?
(2) A rectangle PQRS is 3 times as long and 3 times as wide as another rectangle ABCD. How many times would ABCD fit into PQRS?
(3) A triangle has a base 5 cm long and a height of 6 cm. If it is enlarged so that the base becomes 12.5 cm, what is the new height?
(4) A kite has diagonals 8 cm and 3 cm. If it is enlarged so that the shorter diagonal becomes 9 cm, what is the new length of the large diagonal?
(5) A square is enlarged so that the diagonal becomes four times as long. How much larger is the area?
(6) A triangle is enlarged so that the area becomes 25 times as large. What is the magnification of the triangle?
(7) A triangle with angles 50°, 60° and 70° is enlarged $2\frac{1}{2}$ times. What are the angles of the enlarged triangle?
(8) A rectangle is 3 times as long as it is wide. If it is magnified 5 times, what is the ratio of length : breadth in the enlarged rectangle?
(9) A regular octagon has an area of 20 cm². If it is enlarged so that the length of the sides is trebled, what is the new area?
(10) $\triangle$ABC and $\triangle$PQR are similar. AB : BC = 3 : 2 and AB : PQ = 1 : 2. If AB = 15 cm, calculate QR.
(11) The radii of two spheres are in the ratio 6 : 1. What is the ratio of their surface areas?
(12) The lengths of the sides of two cubes are in the ratio 3 : 1. What is the ratio of their volumes?
(13) Two solid metal spheres are made of the same material. If one of them is twice as wide as the other, what is the ratio of their weights?
(14) A house is 10 m high and a model of the house is 10 cm high. If the ground area of the model is 70 cm², what is the ground area of the house?
(15) A model of one of the Egyptian pyramids is made with 1 kg of plaster. Another model is made which is twice as tall. How much plaster will be needed?
(16) A rectangle has length 10 cm and width 7 cm. If it is sheared to form a parallelogram, what is the area of the parallelogram?
(17) What shape is obtained if a circle is sheared?
(18) A regular hexagon is sheared. Do opposite sides stay parallel?

(19) Two parallel sides of a parallelogram are each 9 cm long and they are 3 cm apart. What is the area of the parallelogram?

(20) A triangle with $a = 3$ cm, $b = 6$ cm, $\angle C = 90°$ is sheared so that it becomes an isosceles triangle with b as base. What is the area of the isosceles triangle?

Exercise 43b

(1) A house is 8 m long and 6 m wide. A scale model of the house is 8 cm long. Calculate

(a) the width of the model

(b) the height of the house if the model is 9 cm tall

(c) the ground area of the model

(d) the volume of the house in m³ if the model has a volume of 900 cm³.

(2) Two similar cones have heights of 4 cm and 12 cm. Calculate

(a) the radius of the small cone if the large cone has radius 9 cm

(b) the area of the curved surface of the large cone if it is 50 cm² for the small cone

(c) the area of the base of the small cone if it is 270 cm² for the large cone

(d) the volume of the large cone if the volume of the small cone is 38 cm³.

(3) Two similar square pyramids have base areas of 9 cm² and 36 cm² respectively. Calculate:

(a) the height of the large pyramid if the small one is 9 cm tall

(b) the ratio of height to width for the small pyramid if it is 3 : 1 for the large one

(c) the angle of slope of the edges of the large pyramid if it is 76° 45′ for the small one

(d) the volume of the large pyramid if the volume of the small one is 27 cm³.

(4) A hollow bowl is in the shape of a hemisphere (half a sphere), of radius 10 cm. A similar bowl has a radius of 20 cm.

(a) Calculate the circumference of the large bowl if the circumference of the small bowl is 64 cm.

(b) Calculate the surface area of the large bowl if the surface area of the small bowl is 629 cm².

(c) Calculate how much liquid the large bowl will hold, if the small bowl holds 2.1 litres.

Section H
44 Quadratic Graph

Quadratic Function

If $y = ax^2 + bx + c$, where a, b and c are constants and can have any value, y is a *quadratic function* of x. If we make a table of values for a quadratic function and plot the points on co-ordinate axes, a smooth curve can be drawn through the points and the result is a *quadratic graph*.

If the value of a is positive, the graph has a lowest point, called a *minimum point*. If the value of a is negative, the graph has a highest point, called a *maximum point*. The maximum or minimum point of the graph is the *turning point*.

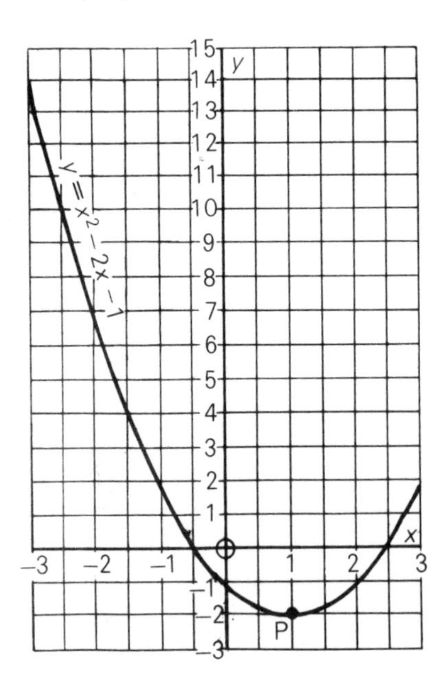

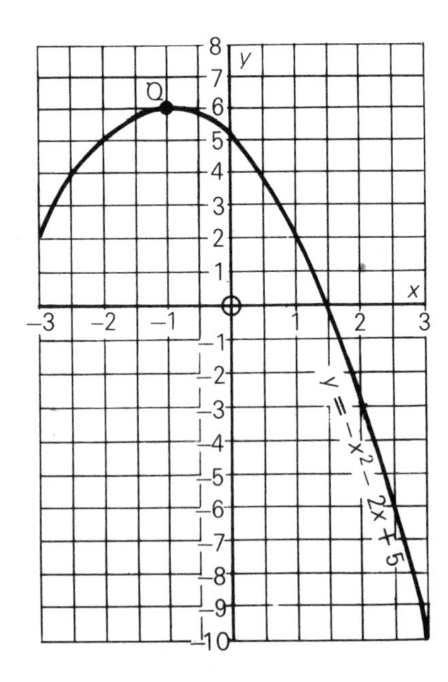

On the graphs shown above, P is the minimum point of one graph, and Q is the maximum point of the other.

Drawing the Graph

The table of values is drawn up as shown below. The table is for the graph of $y = x^2 - 2x - 1$.

x	-3	-2	-1	0	1	2	3
x^2	9	4	1	0	1	4	9
$-2x$	6	4	2	0	-2	-4	-6
-1	-1	-1	-1	-1	-1	-1	-1
y	14	7	2	-1	-2	-1	2

The value of y is found by adding the three terms of the function. The double lines are to remind you not to add the value of x. Notice that the table of values shows which is the minimum point of the graph. It is the point $(1, -2)$. The graph is drawn on page 206.

Sometimes it is not easy to draw the lowest point of the graph accurately and an extra point must be plotted. The table of values given below is for the graph of $y = x^2 - x - 1$.

x	-3	-2	-1	0	1	2	3	$\frac{1}{2}$
x^2	9	4	1	0	1	4	9	$\frac{1}{4}$
$-x$	3	2	1	0	-1	-2	-3	$-\frac{1}{2}$
-1	-1	-1	-1	-1	-1	-1	-1	-1
y	11	5	1	-1	-1	1	5	$-1\frac{1}{4}$

The smallest value of y is -1 at the points where $x = 0$ and $x = 1$. It seems from the table as though the minimum point is between $x = 0$ and $x = 1$. An extra point is therefore calculated for $x = \frac{1}{2}$. This gives the point $(\frac{1}{2}, -1\frac{1}{4})$ which is the minimum point.

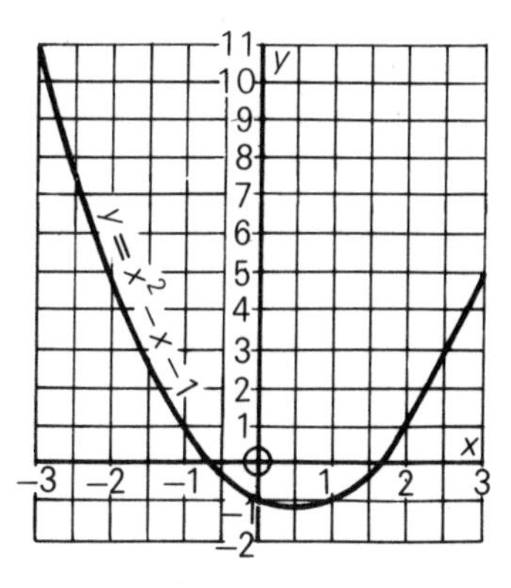

The graph goes from -3 to $+3$ on the x-axis and from $-1\frac{1}{4}$ to $+11$ on the y-axis. Scales of 2 cm to 1 unit of x and 1 cm to 1 unit of y are probably the best, but this will depend on the size of your graph paper.

Finding the Gradient

If a straight line touches a quadratic graph but does not cut it, then it is a *tangent* to the curve. When drawing a tangent to a circle, we know that it is perpendicular to the radius at the point of contact. This helps us to draw the tangent.

There is no easy way of drawing a tangent to a quadratic graph. You simply have to place your ruler so that you can draw a line which touches the graph, but does not cut it. In the figure, AB is a tangent to the curve. The point of contact with the curve is (2, 3). The x co-ordinate of the point of contact is 2, so AB is the tangent at $x = 2$.

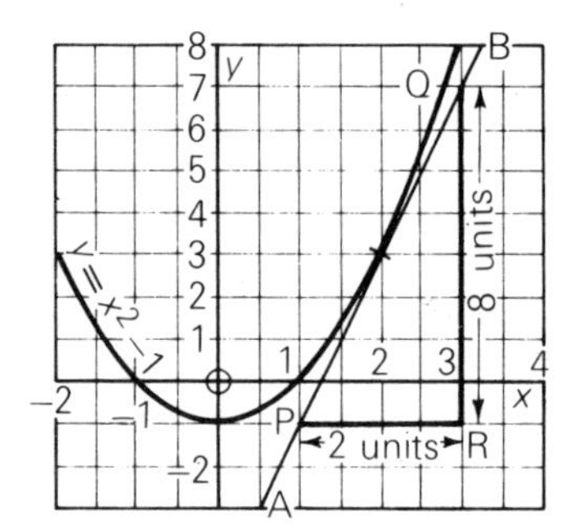

The gradient of a straight line is the same at all points on the line. The gradient of a quadratic graph changes from point to point. The gradient of the graph at any point is the same as the gradient of the tangent at that point. The gradient of the tangent AB is

$$\frac{QR}{PR} = \frac{8}{2} = 4$$

so the gradient of the curve at $x = 2$ is 4.

Exercise 44a

(1) Calculate the value of $2x^2 - 3x + 4$ if x is (i) 1 (ii) 0 (iii) -1.
(2) Calculate the value of $3x^2 + 2x - 1$ if x is (i) 2 (ii) -2 (iii) -1.
(3) Calculate the value of $x^2 - 7x + 1$ if x is (i) 0 (ii) 2 (iii) -2.
(4) Calculate the value of $2x^2 - x + 5$ if x is (i) -1 (ii) 3 (iii) 2.
(5) Calculate the value of $4x^2 - 2x + 7$ if x is (i) 0 (ii) -3 (iii) $\frac{1}{2}$.

Copy and complete the following tables:

(6)

x	-3	-2	-1	0	1	2	3	$-\frac{1}{2}$
$2x^2$	18							
$2x$		-4		0				
-1						-1		-1
$y = 2x^2 + 2x - 1$							23	$-1\frac{1}{2}$

(7)

x	-3	-2	-1	0	1	2	3	$\frac{1}{2}$
$-2x^2$		-8			-2			
$2x$			-2				6	1
3	3					3		
$y = -2x^2 + 2x + 3$				3				$3\frac{1}{2}$

In questions 8–10, plot the given points and draw a smooth curve through them. In each case use a scale of 2 cm to 1 unit of x and 1 cm to 1 unit of y.

(8)

x	-3	-2	-1	0	1	2	$-\frac{1}{2}$
y	10	2	-2	-2	2	10	$2\frac{3}{4}$

(9)

x	-2	-1	0	1	2	3	$\frac{1}{2}$
y	-10	-2	2	2	-2	-10	$2\frac{1}{2}$

(10)

x	-3	-2	-1	0	1	2	$-\frac{1}{2}$
y	-13	-1	5	5	-1	-13	$5\frac{3}{4}$

Draw the graphs of each of the following functions, and write down the co-ordinates of the turning point in each case.

(11) $y = x^2 - 2x - 1$ from $x = -3$ to $x = +4$.
(12) $y = x^2 + 2x + 3$ from $x = -3$ to $x = +3$.
(13) $y = 2x^2 + 4x - 3$ from $x = -4$ to $x = +2$.
(14) $y = -2x^2 - 4x - 3$ from $x = -2$ to $x = +4$.
(15) $y = -x^2 + 5$ from $x = -4$ to $x = +4$.
(16) $y = x^2 - x + 1$ from $x = -3$ to $x = +4$.

(17) $y = 5x^2 + 10x - 5$ from $x = -3$ to $x = +3$.
(18) $y = 4x^2 - 4x + 3$ from $x = -4$ to $x = +4$.
(19) $y = -2x^2 + 2x - 3$ from $x = -6$ to $x = +6$.
(20) $y = -x^2 + 2x - 5$ from $x = -5$ to $x = +5$.

Exercise 44b

(1) (a) Make a table of values for the function $y = x^2$ from $x = -4$ to $x = +4$.
(b) Draw a graph of the function from $y = -4$ to $y = +4$.
(c) Use your graph to find the area of a square with sides 3.5 cm.
(d) Use your graph to find the length of the side of a square whose area is 10 cm^2.
(2) (a) Make a table of values for the function $y = x^2 - 2x - 3$ from $x = -4$ to $x = +4$.
(b) Draw the graph of the function from $x = -4$ to $x = +3$.
(c) Write down the co-ordinates of the turning point and say whether it is a maximum or a minimum point.
(d) By drawing a tangent to your graph, calculate the gradient of the graph at $x = 2$.
(3) (a) Make a table of values for the function $y = -x^2 - 2x + 1$ from $x = -4$ to $x = +4$.
(b) Draw the graph of the function from $x = -4$ to $x = +3$.
(c) Write down the co-ordinates of the turning point and say whether it is a maximum or a minimum point.
(d) Find the gradient of the curve at $x = \frac{1}{2}$.
(4) (a) Make a table of values for the function $y = x^2 + x - 5$ from $x = -5$ to $x = +5$.
(b) Draw the graph of the function from $x = -4$ to $x = +5$.
(c) Write down the co-ordinates of the turning point and say whether it is a maximum or a minimum point.
(d) Find the gradient of the curve at $x = -1$.
(5) (a) Make a table of values for the function $y = -x^2 - x + 3$ from $x = -5$ to $x = +5$.
(b) Draw the graph of the function from $x = -3$ to $x = +4$.
(c) Find the gradient of the graph at $x = 1$.
(d) Find the gradient of the graph at $x = -2$.

45 Length, Area and Volume

Plane Shapes

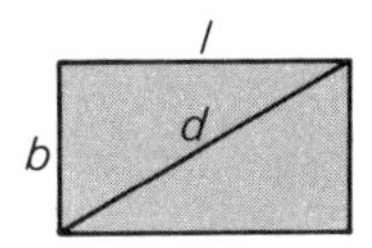

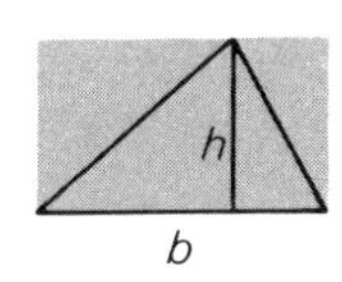

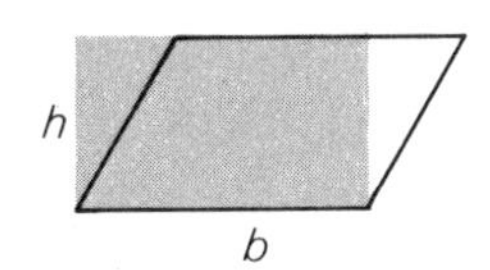

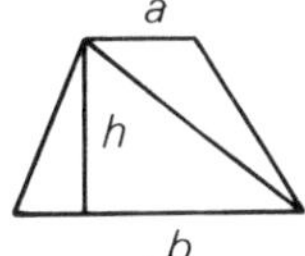

The area of a rectangle is length × breadth, or $A = lb$. The *perimeter* of the rectangle is $2l + 2b = 2(l + b)$. The diagonal d can be found by using the Pythagoras rule, $d^2 = l^2 + b^2$.

If the base of a triangle is b and its height is h, its area is half the area of a rectangle of length b and width h. Therefore, the area of a triangle is $\frac{1}{2}$ base × height, or $\frac{1}{2}bh$.

The parallelogram shown above can be obtained by shearing the rectangle. The area of the parallelogram is therefore $A = bh$, where b is the length of one pair of sides and h is the distance between them.

The area of the trapezium also depends on the length of the parallel sides and on the distance between them. In the figure shown above, the trapezium is made up of two triangles. One has base b and height h. The other has base a and height h. Therefore the area of the trapezium is $\frac{1}{2}ah + \frac{1}{2}bh = \frac{1}{2}(a + b)h$.

The circumference of a circle radius r is $2\pi r$ and the area of the circle is πr^2. The shaded *annulus* between the two circles has area

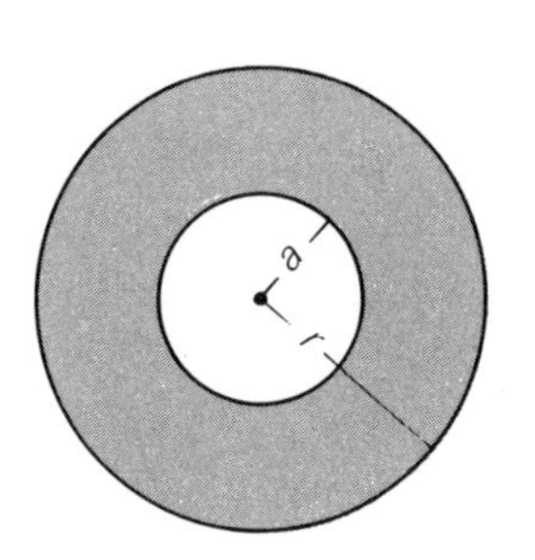

$$\begin{aligned}\pi r^2 - \pi a^2 &= \pi(r^2 - a^2)\\ &= \pi(r + a)(r - a).\end{aligned}$$

Solids

A prism is a solid with a constant cross-section. A prism with a triangular cross-section is a *triangular prism*. A prism with a rectangular cross-section is a *cuboid*. A prism with a circular cross-section is a *cylinder*.

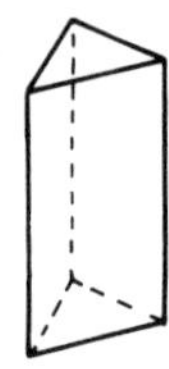
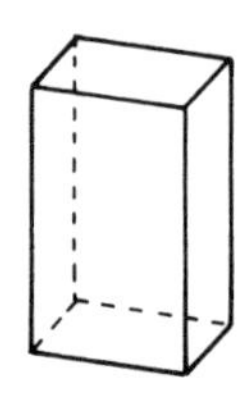
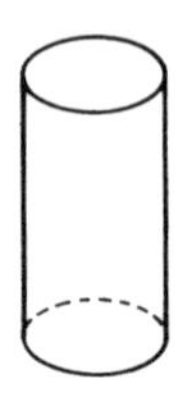

If the cross-section of a prism has an area A and if the height of the prism is h, then the volume is $V = Ah$ no matter what the shape of the cross-section is.

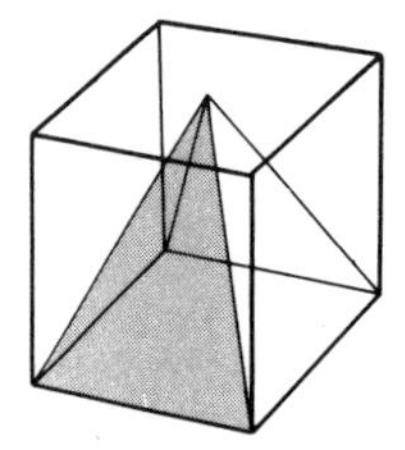
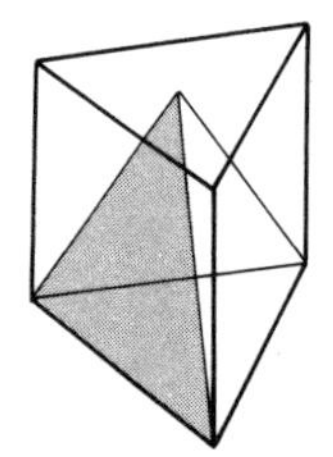
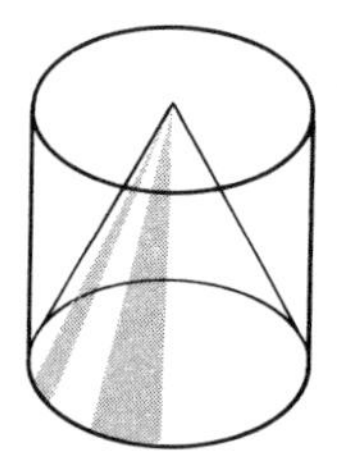

A pyramid fits inside a prism. A *triangular pyramid* fits inside a triangular prism. A *square pyramid* fits inside a square prism. A *cone* fits inside a *cylinder*. The volume of a pyramid is $\frac{1}{3}$ of the volume of the corresponding prism. Thus, if the area of the base is A and the height is h, the volume is $V = \frac{1}{3}Ah$. For a cone the volume is $\frac{1}{3}\pi r^2 h$.

The volume of a sphere of radius r can be shown to be $V = \dfrac{4\pi r^3}{3}$.

Surface Area of Solids

For solids with no curved surfaces, the surface area can be found for each face separately. The area of the curved surfaces of the cylinder and cone can be found by looking at the net of each of these solids.

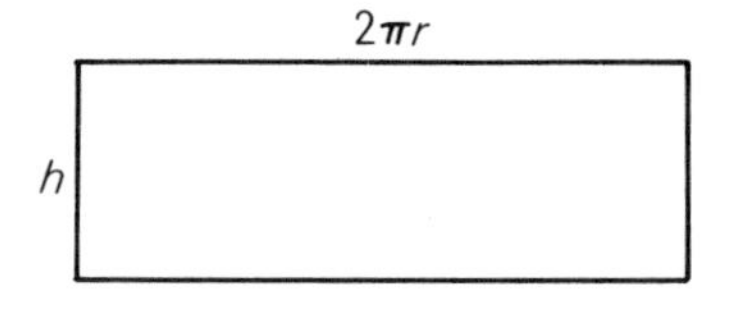

The curved surface of a cylinder opens out to form a rectangle of length $2\pi r$ and width h. Its curved surface area is therefore $S = 2\pi rh$.

The curved surface of a cone opens out to form a sector of a circle. The radius of this sector is the slant height of the cone and the length of the arc is equal to the circumference of the base of the cone. The area of the curved surface can be shown to be $S = \pi rl$. If the height of the cone and the radius are known, the slant height can be found by using the Pythagoras rule, giving $l^2 = r^2 + h^2$.

The surface area of a sphere can be shown to be $S = 4\pi r^2$.

Formulae

Below is given a list of formulae for areas and volumes, and where possible for the perimeters of plane shapes. You will not be expected to remember these formulae, but you must be able to use them and to know what the symbols mean.

plane shape	perimeter	area
rectangle	$P = 2(l + b)$	$A = lb$
parallelogram		$A = bh$
triangle	$P = a + b + c$	$A = \frac{1}{2}bh$
trapezium		$A = \frac{1}{2}(a + b)h$
circle	$C = 2\pi r$	$A = \pi r^2$

solid	volume	area of curved surface
prism	$V = Ah$	
pyramid	$V = \frac{1}{3}Ah$	
cylinder	$V = \pi r^2 h$	$S = 2\pi rh$
cone	$V = \frac{1}{3}\pi r^2 h$	$S = \pi rl$
sphere	$V = \frac{4}{3}\pi r^2 h$	$S = 4\pi r^2$

Exercise 45a

Use logarithms where necessary, and give your answers correct to 3 significant figures; take $\pi = 3.142$.

(1) Calculate the perimeter of a rectangle of length 6.3 cm and width 5.4 cm.

(2) Calculate the perimeter of $\triangle ABC$ in which AB = 7.3 cm, BC = 5.1 cm and AC = 8.2 cm.
(3) Calculate the circumference of a circle radius 3.8 cm.
(4) Calculate the circumference of a circle radius 8.4 cm.
(5) Calculate the area of a rectangle of length 25.2 cm and width 3.7 cm.
(6) Calculate the area of a rectangle of length 7.8 cm and width 9.2 cm.
(7) A parallelogram ABCD has AB = CD = 11.3 cm. If the distance between AB and CD is 3.7 cm, calculate the area of the parallelogram.
(8) A parallelogram PQRS has QR = PS = 2.1 cm. If the distance between QR and PS is 19.3 cm, calculate the area of PQRS.
(9) A triangle has base 4.3 cm and height 6.1 cm. Calculate the area of the triangle.
(10) The two shorter sides of a right-angled triangle are 7.9 cm and 11.5 cm. Calculate the area of the triangle.
(11) If the length and breadth of a rectangle are 7.3 cm and 9.2 cm, calculate the length of the diagonal.
(12) Calculate the diagonal of a square of side 16.7 cm.
(13) Calculate the area of a square of side 7.2 cm.
(14) Calculate the length of the side of a square if the area is 57.6 cm^2.
(15) The parallel sides of a trapezium are 7.3 cm and 11.5 cm. Calculate the area, if the distance between the parallel sides is 6.2 cm.
(16) ABCD is a quadrilateral with $AB \parallel CD$. AB = 3.7 cm and CD = 8.3 cm. If the distance between AB and CD is 5.7 cm, calculate the area of the quadrilateral.
(17) Calculate the area of a circle with radius 4.2 cm.
(18) Calculate the area of a circle with radius 7.3 cm.
(19) Calculate the volume of a cuboid if it has length 18.9 cm, width 7.5 cm and height 4.9 cm.
(20) A prism has a cross-sectional area of 42.5 cm^2 and is 13.5 cm long. Calculate the volume of the prism.
(21) The base of a pyramid has an area of 37.2 cm^2. If the height of the pyramid is 8 cm, calculate its volume.
(22) A pyramid has a square base of side 4.5 cm. If the height of the pyramid is 12.5 cm, calculate its volume.
(23) Calculate the curved surface area of a cylinder with radius 7.3 cm and height 0.3 cm.
(24) Calculate the total surface area of a cylinder with radius 3.7 cm and height 7.9 cm.
(25) If the base radius of a cone is 3.5 cm and the height is 9.6 cm, use the Pythagoras rule to calculate the slant height.

(26) Calculate the curved surface area of a cone with base radius 3.2 cm and slant height 5.7 cm.

(27) Calculate the volume of a cone with radius 4.6 cm and height 7.2 cm.

(28) Calculate the surface area of a sphere of radius 2.7 cm.

(29) Calculate the total surface area of a hemisphere of radius 3.2 cm.

(30) Calculate the volume of a sphere of radius 7.6 cm.

Exercise 45b

(1) A cube has sides of length 5.3 cm. Calculate
 (a) the total surface area
 (b) the length of the diagonal of a face
 (c) the volume.

(2) A cone has base radius 4.2 cm and height 7.7 cm. Calculate
 (a) the slant height
 (b) the area of the curved surface
 (c) the area of the base
 (d) the volume.

(3) A cylinder has radius 6.8 cm and height 9.7 cm. Calculate
 (a) the area of the curved surface
 (b) the total surface area
 (c) the volume.

(4) A parallelogram ABCD has AB = 7.3 cm, BC = 5.2 cm and the distance between AB and CD is 3.5 cm.
 (a) Calculate the area of ABCD.
 (b) Calculate the distance between AD and BC.
 (c) A point P is marked on AB so that AP = 3.9 cm. Calculate the area of APCD.

(5) A hemisphere has a radius of 5.5 cm. Calculate
 (a) the area of the curved surface
 (b) the total surface area
 (c) the volume
 (d) the volume of the cone which has the same base radius and the same height as the hemisphere.

46 Similar Triangles

Sides and Angles

If two triangles have the same shape they are said to be *similar*. The three angles of the first triangle are equal to the three angles of the second triangle.

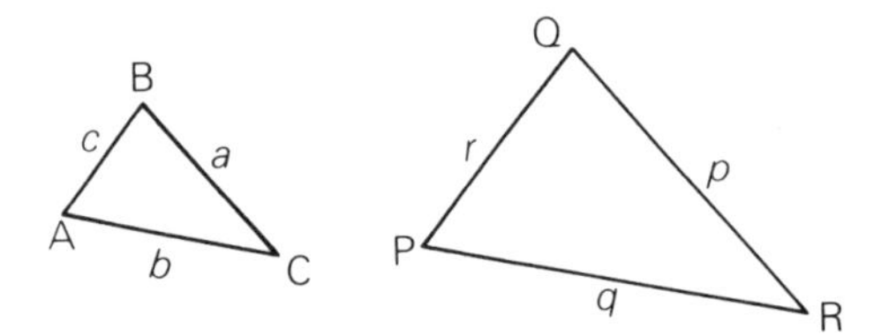

$\triangle$'s, ABC and PQR are similar. $\angle A = \angle P$, $\angle B = \angle Q$ and $\angle C = \angle R$.

AB in the small triangle corresponds to PQ in the large triangle. AB and PQ are *corresponding sides*. The ratio of corresponding sides is

$a:p = b:q = c:r.$

Because the ratio is the same for each pair of corresponding sides,

$\frac{a}{p} = \frac{b}{q} = \frac{c}{r}.$

This also tells us that $\frac{a}{b} = \frac{p}{q}$, $\frac{a}{c} = \frac{p}{r}$, and so on. If the ratio of corresponding sides is $n:1$, the ratio of corresponding areas is $n^2:1$.

Conditions for Similarity

Two triangles are known to be similar if all pairs of corresponding sides are in the same ratio. They are also known to be similar if the three angles of the first triangle are equal to the three angles of the second triangle.

Example: AB and CD are two chords of a circle which intersect at X. Show that $\triangle$'s AXC and DXB are similar, and that $AX \cdot XB = CX \cdot XD$.

In $\triangle$'s AXC, DXB

$\angle A = \angle D$ ($\angle$'s in the same segment)

$\angle C = \angle B$ ($\angle$'s in the same segment)

$\angle AXC = \angle BXD$ (vertically opposite $\angle$'s)

$\therefore$ $\triangle$'s AXC, DXB are similar

and $\dfrac{AX}{XD} = \dfrac{CX}{XB}$.

Multiply both sides by XD and by XB, then

$\therefore AX \cdot XB = CX \cdot XD$.

The above result is an important one. If the two chords do not intersect inside the circle, then if they are not parallel they will intersect outside, and the result is still true in this case.

Congruent Triangles

If two triangles are similar and the ratio of corresponding sides is 1 : 1 the triangles are said to be *congruent*. They are the same shape and size. There are four ways to test whether or not two triangles are congruent.

1. Three sides of one triangle are equal to three sides of the other (SSS).

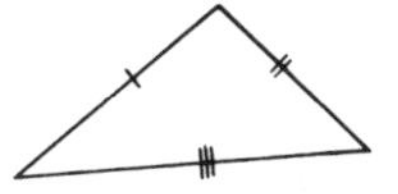
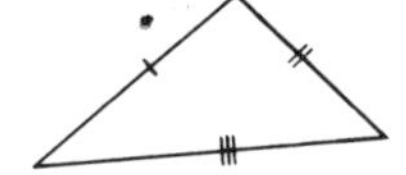

2. Two sides of one triangle are equal to two sides of the other and the included angles are equal (SAS).

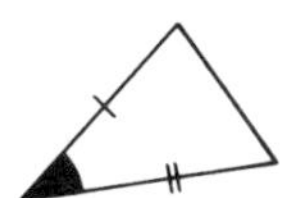
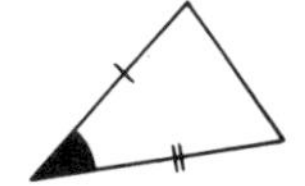

3. Two angles of one triangle are equal to two angles of the other and one pair of corresponding sides is equal (ASA).

4. For right-angled triangles, the hypotenuse and another side of one triangle are equal to the hypotenuse and one other side of the other (RHS).

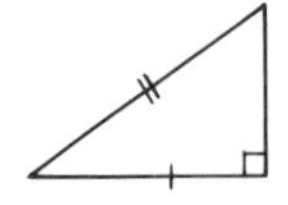
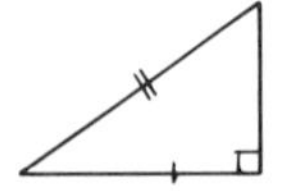

Example: AB is a chord of a circle with centre O. The perpendicular from O to AB cuts AB at X. Show that △'s OXA and OXB are congruent and that AX = XB.

In △'s OXA and OXB
∠OXA = ∠OXB = 90° (OX ⊥ AB)
OA = OB (equal radii)
OX is in both triangles
∴ △'s OXA and OXB are congruent (RHS).

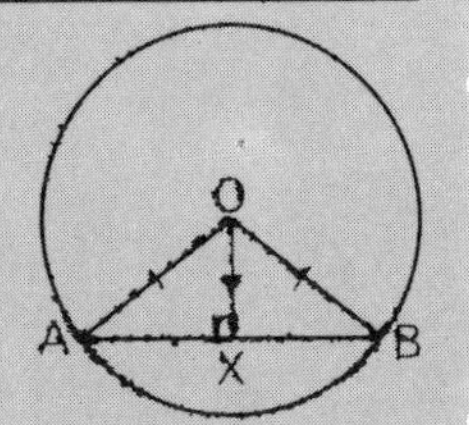

Because the triangles are congruent, corresponding sides are equal, so that AX = XB.

Triangles which are congruent are equal in area.

Exercise 46a

(1) In △'s ABC and PQR, ∠A = ∠P and ∠B = ∠Q. Show that ∠C = ∠R.

(2) Give reasons to show why the named triangles are similar.

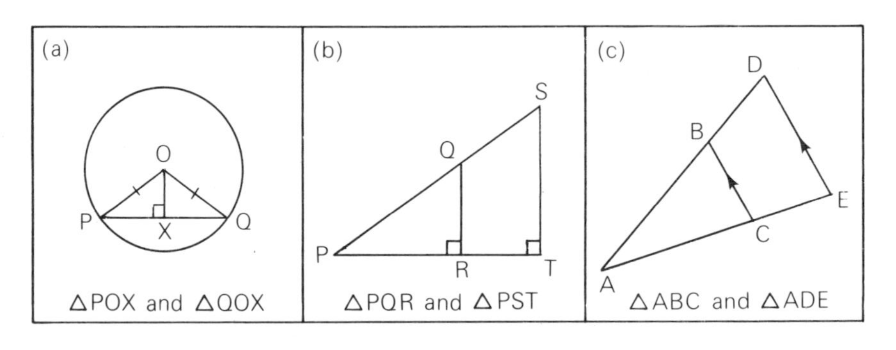

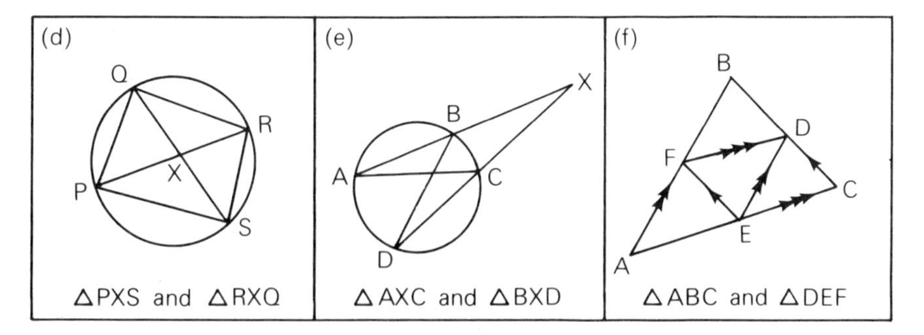

(3) Find the ratio of the areas of △PQR and △XYZ if the ratio of corresponding sides is (i) 2 : 1 (ii) 3 : 1 (iii) 1 : 5 (iv) 1 : 4.

(4) Find the ratio of corresponding sides in △ABC and △DEF if the ratio of the areas is (i) 9 : 1 (ii) 25 : 1 (iii) 1 : 16 (iv) 1 : 64.

(5) Give reasons to show why the named triangles are congruent.

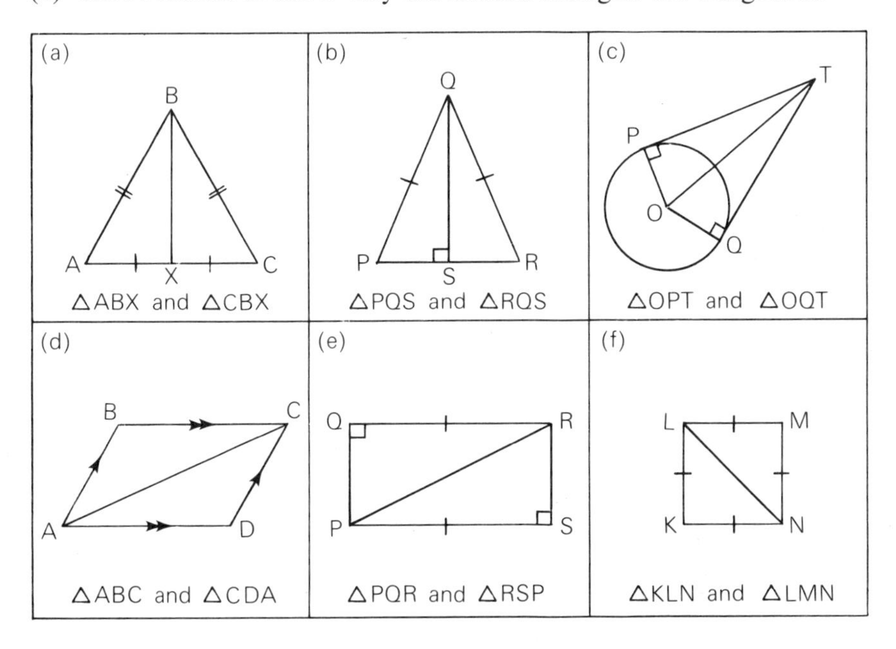

Exercise 46b

(1) In △ABC, AB = 4 cm, BC = 6 cm and AC = 10 cm. D, E and F are the mid-points of AB, BC and CA respectively.

(a) Draw △ABC as accurately as you can.

(b) Name two congruent triangles and show how you know that they are congruent.

(c) Name two triangles which are similar but not congruent, and show how you know that they are similar.

(d) What is the ratio of areas △ABC : △DEF?

(2) Draw $\triangle ABC$ with $AB = 8$ cm, $BC = 6$ cm and $\angle B = 90°$.

(a) Construct a perpendicular from B to AC, and mark the foot of the perpendicular as X. Mark BX as h. Use the Pythagoras rule to calculate the length of AC.

(b) Show that $\triangle ABX$ and $\triangle CAB$ are similar.

(c) Write down the ratio of corresponding sides of $\triangle ABX$ and $\triangle CAB$.

(d) Calculate the ratio of the areas of $\triangle ABX$ and $\triangle CAB$.

(3) In $\triangle ABC$, $a = 5$ cm, $b = 6$ cm and $\angle C = 49°$. If $\triangle ABC$ is enlarged to $\triangle A'B'C'$ where a and a', b and b', and c and c' are pairs of corresponding sides, with $a' = \frac{3a}{2}$,

(a) write down the size of $\angle C'$

(b) calculate the length of b'

(c) calculate the ratio of areas of $\triangle ABC$ and $\triangle A'B'C'$.

(d) Find the ratio $b' : a'$.

(4) Tangents are drawn from a point P to a circle centre O and the points of contact are marked A and B.

(a) Show that $\triangle OAP$ and $\triangle OBP$ are congruent.

(b) If OP and AB intersect at X, show that $\triangle OXA$ and $\triangle OXB$ are congruent.

(c) Show that $\triangle OXB$ and $\triangle OBP$ are similar.

(5) AB and CD are any two chords of a circle radius 5 cm which intersect at a point X outside the circle.

(a) Draw the circle. Draw AB and CD and mark the point X. Join AD and BC.

(b) Show that $\triangle ADX$ and $\triangle BCX$ are similar.

(c) Show that $AX \cdot XB = CX \cdot XD$.

47 Matrices

Addition and Subtraction

Two matrices can only be added or subtracted if they have the same number of rows and the same number of columns. For example,

$$\begin{pmatrix} 3 & 0 \\ 2 & 4 \\ 1 & 5 \end{pmatrix} + \begin{pmatrix} 2 & 4 \\ 0 & 3 \\ -1 & 2 \end{pmatrix} = \begin{pmatrix} 5 & 4 \\ 2 & 7 \\ 0 & 7 \end{pmatrix}.$$

In this addition, each of the matrices to be added has 3 rows of figures and two columns, and is a 3×2 matrix. The sum is also a 3×2 matrix. A matrix with only one column is called a *column matrix*.

$$\begin{pmatrix} 3 \\ 2 \end{pmatrix} - \begin{pmatrix} 4 \\ 1 \end{pmatrix} = \begin{pmatrix} -1 \\ 1 \end{pmatrix}.$$

The figures in a matrix are called the *elements* of the matrix. If an element in one matrix is in the same position as an element in another matrix, then the two elements are *corresponding* elements. When adding two matrices, each element in the first matrix is added to the corresponding element in the other matrix. When subtracting two matrices, corresponding elements are subtracted.

Multiplication

When multiplying two matrices, the order of multiplication is important. If **A** and **B** are two matrices **AB** means that **B** is multiplied by **A** and **AB** $\neq$ **BA**. With matrices, as with letters in algebra, it is not necessary to write the multiplication sign.

The multiplication is carried out as shown at the top of the next page.

$$(3 \quad 2)\begin{pmatrix}4\\5\end{pmatrix} = (3 \times 4 + 2 \times 5) = (22)$$

The product is a single element. If the first matrix has two rows, each row multiplies in the same way and each row gives rise to a single element as before

$$\begin{pmatrix}3 & 2\\2 & 1\end{pmatrix}\begin{pmatrix}4\\5\end{pmatrix} = \begin{pmatrix}22\\13\end{pmatrix}.$$

The top row of the first matrix gives the top element of the product. The bottom row of the first matrix gives the bottom element of the product.

Just as two rows in the first matrix give two rows in the product, so two columns in the second matrix give two columns in the product.

$$\begin{pmatrix}3 & 2\\2 & 1\end{pmatrix}\begin{pmatrix}4 & 1\\5 & 3\end{pmatrix} = \begin{pmatrix}22 & 9\\13 & 5\end{pmatrix}.$$

Notice that, as mentioned earlier, if we multiply the two matrices the other way round, the product is different.

$$\begin{pmatrix}4 & 1\\5 & 3\end{pmatrix}\begin{pmatrix}3 & 2\\2 & 1\end{pmatrix} = \begin{pmatrix}14 & 9\\21 & 13\end{pmatrix}.$$

Example: Multiply $\begin{pmatrix}5 & 1\\2 & 0\end{pmatrix}\begin{pmatrix}3 & 4\\2 & 7\end{pmatrix}$.

$$\begin{pmatrix}5 & 1\\2 & 0\end{pmatrix}\begin{pmatrix}3 & 4\\2 & 7\end{pmatrix} = \begin{pmatrix}5 \times 3 + 1 \times 2, & 5 \times 4 + 1 \times 7\\2 \times 3 + 0 \times 2, & 2 \times 4 + 0 \times 7\end{pmatrix}$$

$$= \begin{pmatrix}17 & 27\\6 & 8\end{pmatrix}.$$

Unit Matrix

Consider any 2×2 matrix and multiply it by $\begin{pmatrix}1 & 0\\0 & 1\end{pmatrix}$.

$$\begin{pmatrix}1 & 0\\0 & 1\end{pmatrix}\begin{pmatrix}3 & 7\\5 & 2\end{pmatrix} = \begin{pmatrix}1 \times 3 + 0 \times 5, & 1 \times 7 + 0 \times 2\\0 \times 3 + 1 \times 5, & 0 \times 7 + 1 \times 2\end{pmatrix} = \begin{pmatrix}3 & 7\\5 & 2\end{pmatrix}.$$

Any 2×2 matrix is unchanged when multiplied by $\begin{pmatrix}1 & 0\\0 & 1\end{pmatrix}$ and in this case it does not matter which way round the matrices are multiplied.

$$\begin{pmatrix}3 & 7\\5 & 2\end{pmatrix}\begin{pmatrix}1 & 0\\0 & 1\end{pmatrix}=\begin{pmatrix}3\times 1+7\times 0, & 3\times 0+7\times 1\\5\times 1+2\times 0, & 5\times 0+2\times 1\end{pmatrix}=\begin{pmatrix}3 & 7\\5 & 2\end{pmatrix}.$$

Because this is rather like multiplying a number by 1, the matrix $\begin{pmatrix}1 & 0\\0 & 1\end{pmatrix}$ is called the *unit matrix*.

Transformations

Consider the point P, with x co-ordinate 3 and y co-ordinate 4. This is usually shown as the 1×2 matrix with a comma, (3, 4). Suppose instead that it is represented by the column matrix $\begin{pmatrix}3\\4\end{pmatrix}$.

$$\begin{pmatrix}-1 & 0\\0 & 1\end{pmatrix}\begin{pmatrix}3\\4\end{pmatrix}=\begin{pmatrix}-3\\4\end{pmatrix}.$$

This has the effect of moving P from (3, 4) to $(-3, 4)$. Call this new point P′.

Let $\triangle$PQR have vertices at P(3, 4), Q(4, 4) and R(6, 6). If we multiply the column matrices for Q and R, we get

$$\begin{pmatrix}-1 & 0\\0 & 1\end{pmatrix}\begin{pmatrix}4\\4\end{pmatrix}=\begin{pmatrix}-4\\4\end{pmatrix}$$

$$\begin{pmatrix}-1 & 0\\0 & 1\end{pmatrix}\begin{pmatrix}6\\6\end{pmatrix}=\begin{pmatrix}-6\\6\end{pmatrix}.$$

Thus,

P(3, 4) $\longrightarrow$ P′$(-3, 4)$
Q(4, 4) $\longrightarrow$ Q′$(-4, 4)$
R(6, 6) $\longrightarrow$ R′$(-6, 6)$.

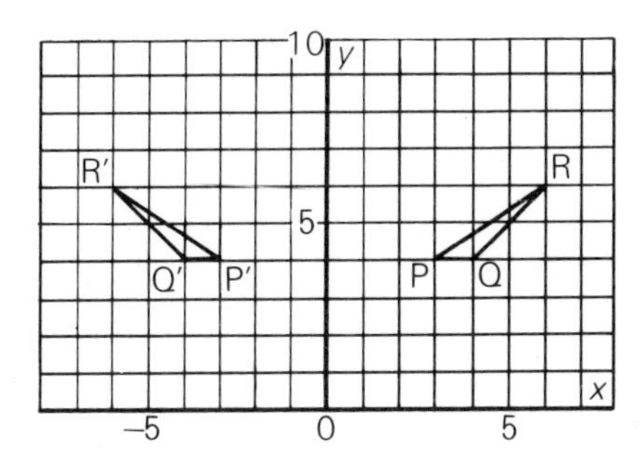

The triangle has been reflected in the y-axis. If multiplying the column matrices by $\begin{pmatrix}-1 & 0\\ 0 & 1\end{pmatrix}$ reflects all points in the y-axis, then by repeating the multiplication, the triangle should be reflected back again to its starting position. The three multiplications can be combined.

$$\begin{pmatrix}-1 & 0\\ 0 & 1\end{pmatrix}\begin{pmatrix}-3 & -4 & -6\\ 4 & 4 & 6\end{pmatrix}=\begin{pmatrix}3 & 4 & 6\\ 4 & 4 & 6\end{pmatrix}.$$

Multiplying twice by $\begin{pmatrix}-1 & 0\\ 0 & 1\end{pmatrix}$ leaves points unchanged. This is because

$$\begin{pmatrix}-1 & 0\\ 0 & 1\end{pmatrix}\begin{pmatrix}-1 & 0\\ 0 & 1\end{pmatrix}=\begin{pmatrix}1 & 0\\ 0 & 1\end{pmatrix}.$$

Although we have only considered the movement of $\triangle$PQR, it can easily be shown that all points are reflected in the y-axis by this matrix multiplication. It is the whole plane which has been reflected.

Example: Plot the points A(−3, −1), B(−5, 5) and O(0, 0). Find the image of $\triangle$ABO when multiplied by $\begin{pmatrix}0 & 1\\ -1 & 0\end{pmatrix}$. Say what kind of transformation this represents.

$$\begin{pmatrix}0 & 1\\ -1 & 0\end{pmatrix}\begin{pmatrix}-3 & -5 & 0\\ -1 & 5 & 0\end{pmatrix}=\begin{pmatrix}-1 & 5 & 0\\ 3 & 5 & 0\end{pmatrix}.$$

A(−3, −1) ⟶ A′(−1, 3)
B(−5, 5) ⟶ B′(5, 5)
O(0, 0) ⟶ O(0, 0).

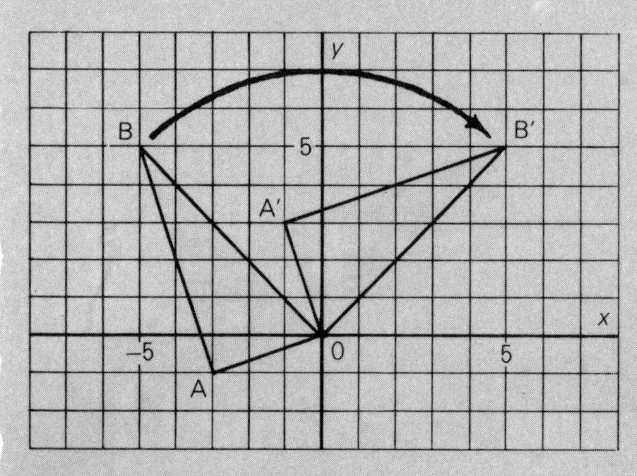

This represents a clockwise rotation of 90° about the origin.

Exercise 47a

Calculate the following:

(1) $\begin{pmatrix} 3 & 2 \\ 4 & 1 \\ -1 & 0 \end{pmatrix} + \begin{pmatrix} 4 & 5 \\ 2 & 3 \\ 6 & 4 \end{pmatrix}$

(2) $\begin{pmatrix} 2 \\ 3 \\ 1 \\ 4 \end{pmatrix} + \begin{pmatrix} 5 \\ -2 \\ 1 \\ 7 \end{pmatrix}$

(3) $\begin{pmatrix} 3 & 4 \\ -1 & 0 \end{pmatrix} + \begin{pmatrix} 2 & -4 \\ 6 & 1 \end{pmatrix}$

(4) $\begin{pmatrix} 6 \\ 3 \end{pmatrix} + \begin{pmatrix} 5 \\ 0 \end{pmatrix}$

(5) $\begin{pmatrix} 8 & 6 & 3 \\ 2 & 1 & -7 \end{pmatrix} + \begin{pmatrix} 1 & 0 & 1 \\ 2 & -1 & 0 \end{pmatrix}$

(6) $\begin{pmatrix} 2 \\ 0 \end{pmatrix} - \begin{pmatrix} 3 \\ 4 \end{pmatrix}$

(7) $\begin{pmatrix} 8 & 1 \\ 4 & 7 \end{pmatrix} - \begin{pmatrix} 3 & 2 \\ 5 & 1 \end{pmatrix}$

(8) $\begin{pmatrix} 4 & 1 \\ 0 & 3 \end{pmatrix} - \begin{pmatrix} 8 & 1 \\ 3 & 1 \end{pmatrix}$

(9) $\begin{pmatrix} 0 & 4 \\ 2 & 1 \end{pmatrix} - \begin{pmatrix} 5 & 2 \\ 1 & 4 \end{pmatrix}$

(10) $\begin{pmatrix} 3 & 4 & 1 \\ 2 & 1 & 0 \end{pmatrix} - \begin{pmatrix} 3 & 3 & 3 \\ 1 & 1 & 1 \end{pmatrix}$

(11) $(1 \quad 3) \begin{pmatrix} 3 \\ 4 \end{pmatrix}$

(12) $(5 \quad 3) \begin{pmatrix} 2 \\ 1 \end{pmatrix}$

(13) $(7 \quad 1) \begin{pmatrix} 1 \\ 2 \end{pmatrix}$

(14) $(6 \quad 3) \begin{pmatrix} 5 \\ 1 \end{pmatrix}$

(15) $\begin{pmatrix} 1 & 3 \\ 6 & 6 \end{pmatrix} \begin{pmatrix} 0 \\ 3 \end{pmatrix}$

(16) $\begin{pmatrix} 5 & 3 \\ 2 & 1 \end{pmatrix} \begin{pmatrix} 2 \\ 2 \end{pmatrix}$

(17) $\begin{pmatrix} 0 & 0 \\ 0 & 0 \end{pmatrix} \begin{pmatrix} 4 \\ 1 \end{pmatrix}$

(18) $\begin{pmatrix} 1 & 0 \\ 0 & 1 \end{pmatrix} \begin{pmatrix} 2 \\ 6 \end{pmatrix}$

(19) $\begin{pmatrix} 3 & 4 \\ 1 & 2 \end{pmatrix} \begin{pmatrix} 5 \\ 3 \end{pmatrix}$

(20) $\begin{pmatrix} 0 & 1 \\ 1 & 0 \end{pmatrix} \begin{pmatrix} 4 \\ 2 \end{pmatrix}$

(21) $\begin{pmatrix} 1 & -1 \\ 2 & 3 \end{pmatrix} \begin{pmatrix} 3 & 1 \\ 1 & 2 \end{pmatrix}$

(22) $\begin{pmatrix} 0 & 3 \\ 1 & -1 \end{pmatrix} \begin{pmatrix} 2 & 1 \\ 4 & 0 \end{pmatrix}$

(23) $\begin{pmatrix} 2 & 0 \\ 0 & 2 \end{pmatrix} \begin{pmatrix} 1 & -3 \\ -1 & 3 \end{pmatrix}$

(24) $\begin{pmatrix} 3 & 0 \\ 0 & 3 \end{pmatrix} \begin{pmatrix} 2 & 1 \\ 4 & 3 \end{pmatrix}$

(25) $\begin{pmatrix} 1 & 1 \\ 1 & 1 \end{pmatrix} \begin{pmatrix} 2 & -3 \\ -2 & 3 \end{pmatrix}$

(26) $\begin{pmatrix} 3 & 5 \\ 1 & 2 \end{pmatrix} \begin{pmatrix} 1 & 0 \\ 2 & 1 \end{pmatrix}$

(27) $\begin{pmatrix} 2 & 5 \\ 1 & 3 \end{pmatrix} \begin{pmatrix} 3 & -5 \\ -1 & 2 \end{pmatrix}$

(28) $\begin{pmatrix} 1 & 2 \\ 1 & 3 \end{pmatrix} \begin{pmatrix} 3 & 2 \\ 1 & 1 \end{pmatrix}$

(29) $\begin{pmatrix} 5 & 1 \\ 1 & 3 \end{pmatrix} \begin{pmatrix} -1 & 2 \\ 1 & 0 \end{pmatrix}$

(30) $\begin{pmatrix} 1 & 2 \\ 2 & -2 \end{pmatrix} \begin{pmatrix} 0 & 1 \\ -3 & 4 \end{pmatrix}$

Exercise 47b

(1) If $\mathbf{A} = \begin{pmatrix} 2 & 3 \\ 1 & 7 \end{pmatrix}$ and $\mathbf{B} = \begin{pmatrix} 5 & 2 \\ 1 & -1 \end{pmatrix}$, find

(a) $2\mathbf{A}$
(b) $\mathbf{A} + \mathbf{B}$
(c) $\mathbf{AB}$
(d) $\mathbf{B}^2$.

(2) Draw co-ordinate axes using a scale of 1 cm to 1 unit on both axes.

(a) Plot the points A(1, 3), B(6, 6) and C(7, 4) and join the points to form △ABC.

(b) Plot the image of △ABC when multiplied by $\mathbf{M} = \begin{pmatrix} -1 & 0 \\ 0 & 1 \end{pmatrix}$.

(c) Name the transformation which has taken place.

(d) By finding $\mathbf{M}^2$, show that a second multiplication by $\mathbf{M}$ would map the triangle back to its original position.

(3) Draw co-ordinate axes using a scale of 1 cm to 1 unit on both axes.

(a) Plot the points D(2, 1), E(−1, 4) and F(−3, −3) and join the points to form △DEF.

(b) Plot the image of △ABC when multiplied by $\mathbf{P} = \begin{pmatrix} 2 & 0 \\ 0 & 2 \end{pmatrix}$.

(c) Name the transformation which has taken place.

(4) Draw co-ordinate axes using a scale of 1 cm to 1 unit on both axes.

(a) Plot the points P(4, 1), Q(6, 1) and R(6, 3) and join the points to form △PQR.

(b) Plot the image of △ABC when multiplied by $\mathbf{A} = \begin{pmatrix} 0 & 1 \\ 1 & 0 \end{pmatrix}$ and name the transformation which has taken place.

(c) Plot the image of △ABC when multiplied by $\mathbf{B} = \begin{pmatrix} 0 & -1 \\ 1 & 0 \end{pmatrix}$ and name the transformation which has taken place.

(5) Draw co-ordinate axes for positive values of x and y, using a scale of 2 cm to 1 unit on both axes.

(a) Plot the points P(1, 0), Q(1, 2), R(2, 0) and S(2, 2). What kind of figure is PQRS?

(b) Plot the image of PQRS when multiplied by $\begin{pmatrix} 1 & 1 \\ 0 & 1 \end{pmatrix}$.

(c) Name the kind of transformation which has taken place.

(d) Find the area of the image of PQRS under this transformation.

48 Plan and Elevation

Looking at Solids

The view of a solid object as seen from above is called the *plan*. The view seen from the front is called the *front elevation* and a view seen from one side is called a *side elevation*. Scale drawings which show the plan and elevations of a solid give us complete information about its shape and size.

The plan and elevations of a solid are always set out in the same way. In order to understand this, imagine a garden shed in mid-air, inside a large cardboard box.

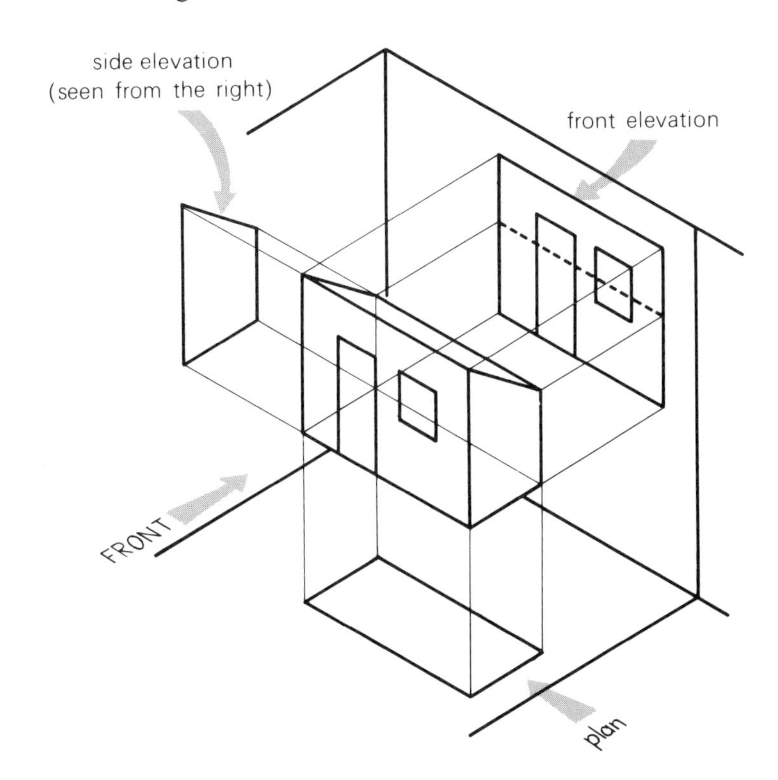

The front view of the shed is drawn at the back of the box, the plan on the bottom of the box and the side views on the sides of the box. In this picture, one side of the box has been cut away so that the other parts can be seen. Notice that the view of the shed as seen from the right, is drawn on the left-hand side of the box.

If we open out the box we see the plan and elevations all together.

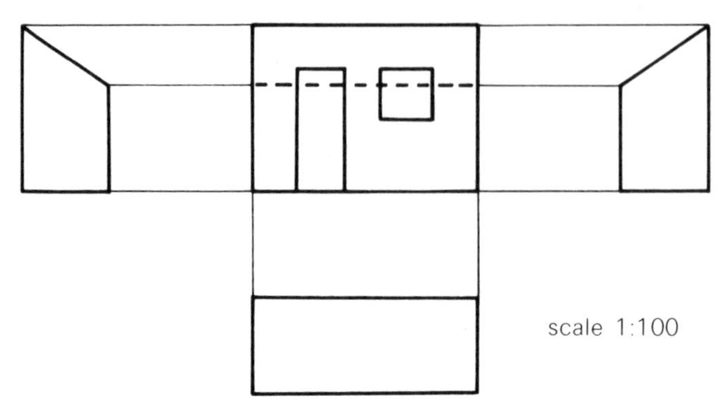

The dotted line shows the hidden edge of the roof.

Drawing Plans and Elevations

Suppose we are to draw a plan and front elevation of the steps shown below and the side elevation as seen from the left.

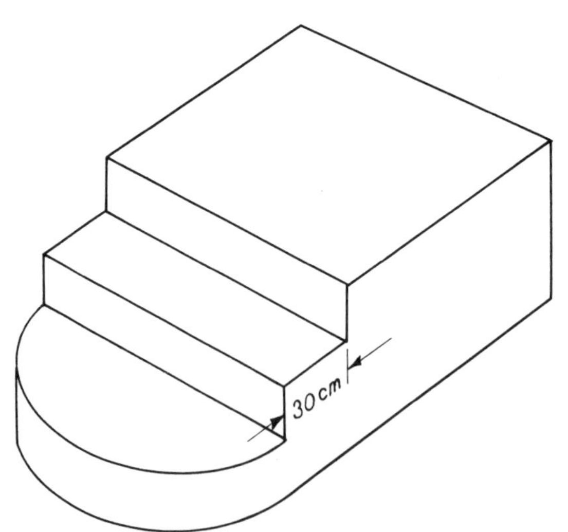

The top step is 1 metre square and the lower step is a semi-circle. Each step is 20 cm high. The plan and elevations can be drawn by following the instructions given overleaf, using a scale of 1 : 20.

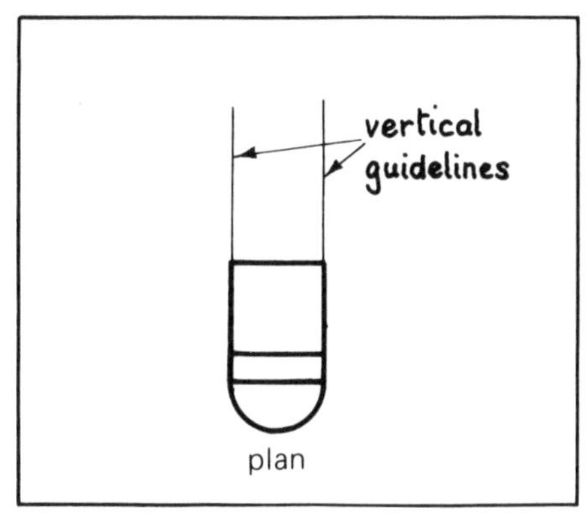

1. Draw the plan and the vertical guidelines for the front elevation.

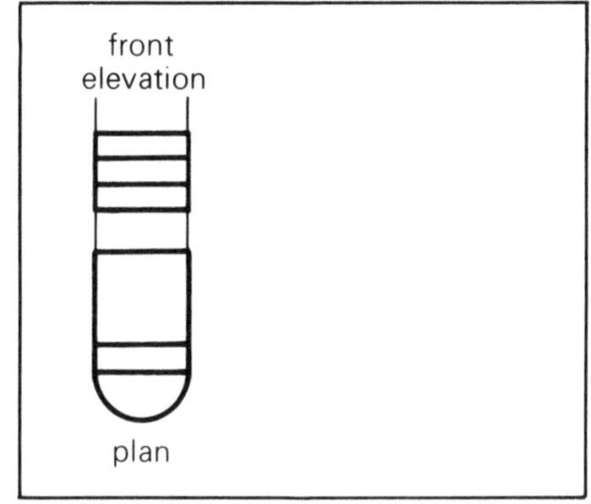

2. Use the guidelines to draw the front elevation.

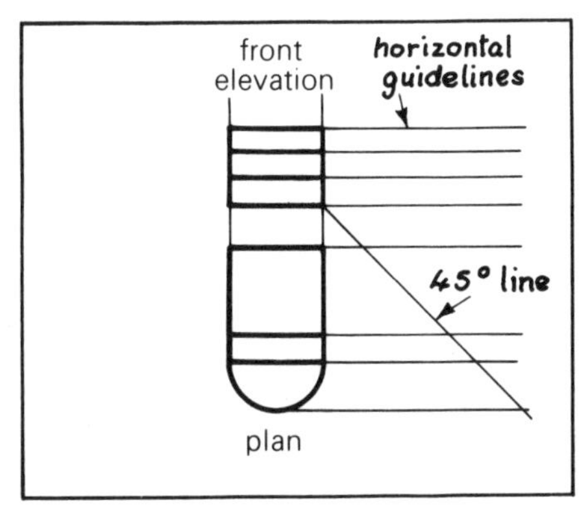

3. Draw all horizontal guidelines and the 45° line.

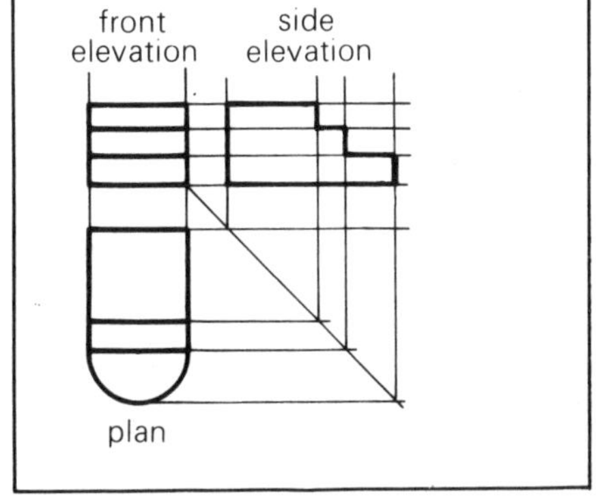

4. Draw vertical guidelines from the 45° line. Use horizontal and vertical guidelines to draw side elevation.

Exercise 48a

(1) Draw a plan, front elevation and one side elevation for each of the shapes shown overleaf. Use any convenient scale, but state clearly what scale you have used.

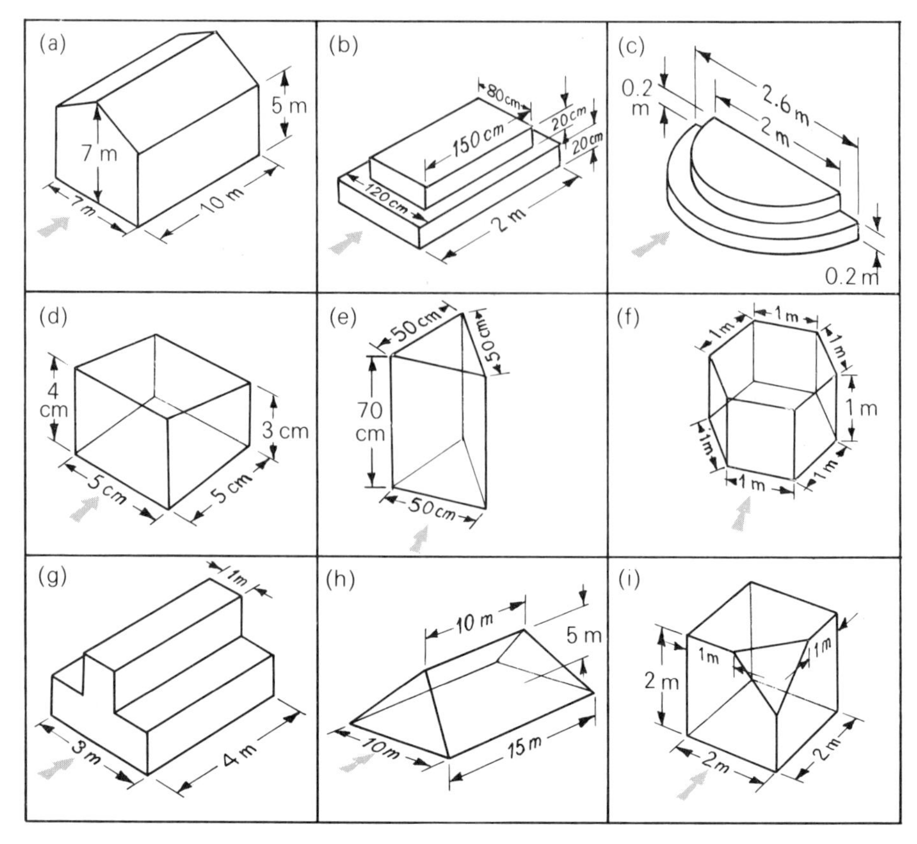

(2) A pyramid has a square base of side 5 cm and its height is 10 cm. The top of the pyramid is cut off parallel to the base and the frustum which remains has a height of 6 cm. Draw a full scale plan and elevation of the frustum.

(3) A clock tower consists of a square prism of side 4 m and height 9 m, with a square pyramid on top. The base of the pyramid fits exactly on to the top of the prism and has a height of 5 m. Draw a plan and elevation of the clock tower using a scale of 1 : 200.

(4) The base of a pyramid is a regular hexagon of side 2 cm and height 7 cm. Draw a full scale plan of the pyramid and the elevation seen from a point directly in front of one of the sloping faces. Draw also the side elevation.

(5) A hurricane lamp is in the shape of a square prism of side 10 cm and height 18 cm. The chimney is a cylinder radius 3 cm standing 2 cm above the top of the prism. Draw the plan and elevation of the lamp to any suitable scale.

(6) A chimney cowl consists of a cone of height 10 cm and base radius 15 cm on top of a cylinder radius 10 cm and height 20 cm. Draw a plan and elevation of the cowl to any suitable scale.

(7) A platform is made of two cylinders. The smaller cylinder has a radius of 1 m and a height of 25 cm. It is placed symmetrically on the top of the larger cylinder which has a radius of 75 cm and the same height as the smaller cylinder. Draw the plan and one elevation, using any suitable scale.

Exercise 48b

(1) The figure shows a vaulting box which is 1 m high.

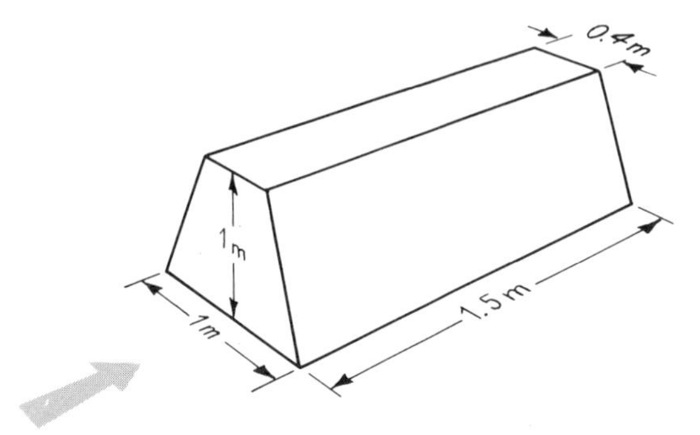

(a) Use a scale of 1 : 20 to make a plan of the box.

(b) Use your plan of the box to draw the front elevation and one side elevation.

(c) From your drawing, find the length of the sloping edge of the box as accurately as possible.

(d) From your drawing, find the angle which the sloping face of the box makes with the horizontal.

(2) The figure shows a kitchen wall unit with a sloping front.

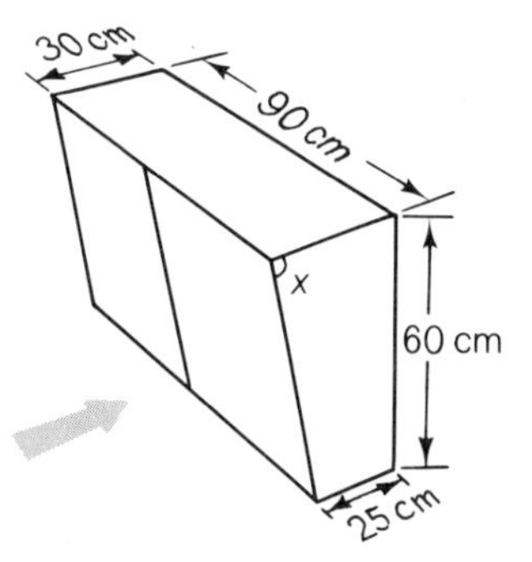

(a) Using a scale of 1 : 10 draw a plan of the unit.

(b) Use your plan to draw the front elevation and the side elevation seen from the left.

(c) From your drawing find the angle marked x to the nearest degree.

(3) The figure shows a model of a house.

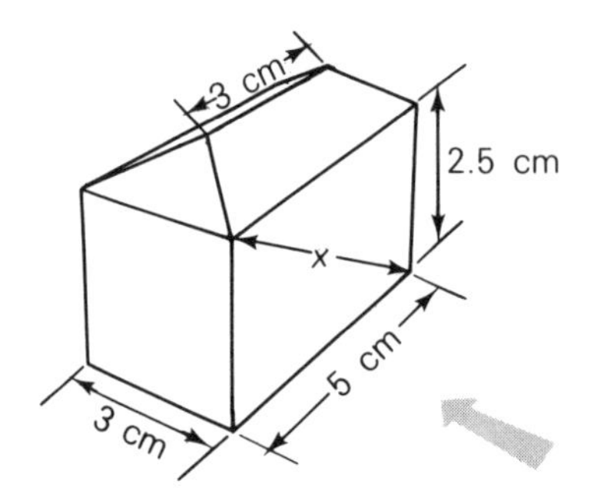

(a) Draw a full scale plan of the model.

(b) Use your plan to draw the front elevation, given that the ridge of the roof is 3.5 cm above the ground.

(c) Draw one side elevation.

(d) From your drawing find the length marked x to the nearest mm.

(4) The figure shows a rectangular block with a hemisphere at one end. The radius of the hemisphere is 3 cm.

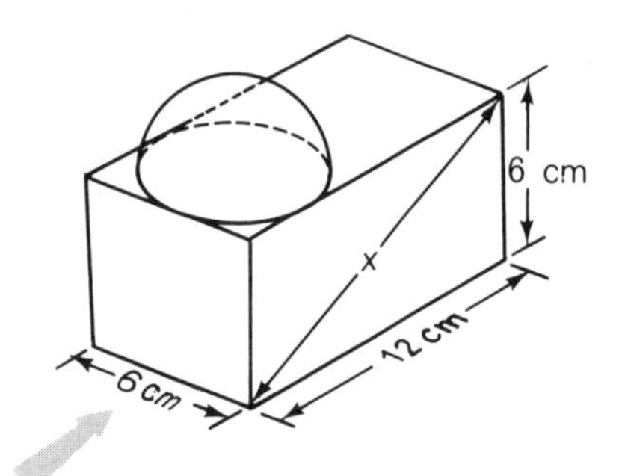

(a) Draw a half-scale plan of the block and hemisphere.

(b) Draw the front elevation.

(c) Draw the side elevation seen from the right.

(d) From your drawing find the length marked x to the nearest mm.

(5) The base of a triangular pyramid is an equilateral triangle of side 25 cm and the height of the pyramid is 20 cm.

(a) Draw the plan of the pyramid using a scale of 1 : 5.

(b) Draw the front elevation of the pyramid (facing a sloping face).

(c) Draw one side elevation of the pyramid.

(d) From your drawing find the angle that a sloping face makes with the horizontal.

(e) From your drawing find the angle that an edge makes with the horizontal.

49 Statistics

Averages

Five boys run a race over a distance of 100 m and their times are recorded as 16.0 s, 16.7 s, 15.3 s, 17.1 s and 15.5 s. The average time for the race is

$$\begin{aligned}\frac{\text{total time}}{\text{number of boys}} &= \frac{16.0 + 16.7 + 15.3 + 17.1 + 15.5}{5}\\ &= \frac{80.6}{5}\\ &= 16.12 \text{ s}\\ &= 16.1 \text{ s to the nearest } \tfrac{1}{10} \text{ of a second.}\end{aligned}$$

We say that 16.1 s is the *mean average* time to the nearest 0.1 s.

Combined Averages

Averages cannot be added directly. Suppose 10 people score an average of 50% in an examination and another 30 people score an average of 40% in the same examination. What is the average mark for all 40 people? The average is not 40% + 50%, nor is it half way between.

If 10 people score an average of 50%, their total score is $10 \times 50 = 500\%$. If 30 people score 40%, their total score is $30 \times 40 = 1\,200\%$. The total score for all 40 people is $500\% + 1\,200\% = 1\,700\%$ and the average mark is

$$\begin{aligned}\frac{\text{total score}}{\text{number of people}} &= \frac{1\,700\%}{40}\\ &= 42.5\%.\end{aligned}$$

Example: In the first quarter of the year, the mean average monthly rainfall at Mexico City was 0.9 cm. For the remainder of the year the mean average monthly rainfall was 8.0 cm. Calculate the mean average monthly rainfall for the whole year, correct to the nearest mm.

Total rainfall for the first 3 months is 3×0.9 cm = 2.7 cm
Total rainfall for the last 9 months is 9×8.0 cm = 72.0 cm
∴ Total rainfall for the whole year is 74.7 cm.

$$\text{Mean average monthly rainfall is } \frac{\text{total rainfall}}{\text{number of months}} = \frac{74.7}{12} = 6.22 = 6.2 \text{ cm.}$$

Probability

Suppose you have a bag with 5 red marbles and 10 green marbles. If you shake up the bag and dip in your hand to take out a marble, you could pick any one of the fifteen marbles. No one marble is more likely to be chosen than any other. There are 15 possible results of your choice and no one result is any more likely than any other result. Because of this, the choice is called a *random event*.

By studying random events we can learn to predict results to some extent and to decide when events are not as random as we think they should be. If, for example, you toss a coin 10 times, you would expect to get 5 heads and 5 tails. If you get 6 heads and 4 tails you would not be surprised. If you get 7 heads and 3 tails you might be suspicious. If there were 10 heads in a row, you would be fairly certain that the coin has a *bias*.

Probability experiments are often done with dice, coins and playing cards. Dice are either in the shape of a cube and numbered from 1 to 6, or in the shape of a regular icosahedron (see chapter 30) and numbered from 1 to 20.

To calculate the chances of an event, we consider that particular result to be a success and calculate the total number of possible results, then

$$\frac{\text{number of ways of getting a success}}{\text{total number of possible results}}$$

is called the *probability* of the event happening. Because the ways of getting a success cannot be more than the total number of possible results, the probability of an event happening can never be greater than 1.

Example: What is the probability of getting a red number at roulette?

There are 37 numbers altogether and 18 of them are red.
∴ Probability of getting a red number is

$$\frac{\text{number of red numbers}}{\text{total number of possible results}} = \frac{18}{37}$$
$$= 0.486$$
$$= 0.49 \text{ (to 2 decimal places).}$$

Exercise 49a

(1) The examination marks of 9 candidates are 67%, 85%, 53%, 42%, 67%, 81%, 94%, 48% and 55%. Calculate the average mark to the nearest whole number.

(2) Seven motorcyclists test the fuel consumption of their bikes by finding how far they can drive on 5 litres of petrol. The results were 85.3 km, 98.7 km, 79.9 km, 102.5 km, 91.6 km, 88.6 km and 83.9 km. Calculate the average fuel consumption in km/l.

(3) A book was opened at random and the number of letters in 50 words were counted. The results were as follows:

2	11	2	2	3	3	7	11	5	2
4	3	2	11	3	10	10	3	8	4
3	7	10	3	5	6	3	7	3	3
4	14	3	2	9	2	9	2	13	3
3	3	9	10	5	3	6	10	6	5

Calculate the mean average number of letters per word correct to 1 decimal place.

(4) The heights of a group of adults was measured to the nearest cm and the results were as follows:

153	172	165	181	154	167	168	154	168	175
169	166	159	171	159	164	171	163	159	179

Calculate the mean average height of the group correct to the nearest cm.

(5) A consumer group bought the same article in a number of different shops and the prices paid were as follows:

£5.37 £6.20 £5.42 £5.35 £5.35
£5.35 £5.35 £5.35 £5.41 £5.00
£5.35 £5.30 £5.35 £5.47 £5.35

Calculate the mean average price paid to the nearest penny.

(6) In a public examination a candidate takes two language papers and three science papers. The average mark for the language papers was 61% and the average for the science papers was 56%. Calculate the average mark for all five papers.

(7) For the first 10 games of the term the school football team scores an average of 1 goal per match, but in the remaining 2 games it scores an average of 7 goals per match. Calculate the average number of goals scored per match for the whole term.

(8) A man drives at an average speed of 31 km/h for 2 hours and then keeps up an average speed of 55 km/h for 4 hours. Calculate his average speed for the whole journey.

(9) In class 3C at Hilltop school there are 15 boys and 20 girls. If the average examination mark for the boys is 43% and for the girls is 48%, calculate the average mark for the whole class, to the nearest whole number.

(10) Five motorcyclists find that their average fuel consumption is 81.3 km/l and 7 of their friends find that their average fuel consumption is 92.4 km/l. Calculate the average fuel consumption for the whole group.

(11) All the children in 5A and 5B were asked to write down the number of children in their family and the results were analysed. 5A has 26 children with a mean average number of 2.5 children per family and 5B has 34 children with an average of 3.0 children per family. Calculate the mean average number of children per family in the two classes, as a whole, correct to 1 decimal place.

(12) The mean average rainfall at Darwin in the first four months of the year is 26.3 cm per month and in the last eight months is 5.6 cm per month. Calculate the average monthly rainfall.

(13) An amateur drama club has 13 women and 11 men. If the average weight of the men is 79.3 kg and the average weight of the women is 57.4 kg, calculate the average weight of all members of the group correct to 3 significant figures.

(14) A sample of 25 light bulbs tested by the manufacturer had an average life of 519 hours. Another sample of 40 bulbs had an average life of 597 hours. Calculate the average life of all the bulbs tested.

(15) A brass alloy is made by mixing 30 cm^3 of copper with 20 cm^3 of tin. If each cm^3 of copper weighs 8.9 g and each cm^3 of tin weighs 7.3 g calculate the average weight of each cm^3 of the alloy.

Calculate the following probabilities for random events:

(16) Throwing a 6 with a 6-sided die.
(17) Picking a club from a pack of playing cards.
(18) Picking a seven from a pack of playing cards.
(19) Scoring a prime number greater than 4 with a 20-sided die.
(20) Getting an even score with a 6-sided die.
(21) Picking a red card from a pack of playing cards.
(22) Picking a picture card from a pack of playing cards.
(23) Scoring a 'square' number with a 20-sided die.
(24) Scoring a 'triangular' number with a 20-sided die.
(25) Picking the 3 of diamonds from a pack of playing cards.

Exercise 49b

(1) Fifty pea pods were opened and were found to contain the following number of peas:

6	5	1	5	6	9	6	3	3	8
7	4	9	8	5	6	8	5	6	6
8	8	7	6	5	5	7	8	6	7
7	6	5	6	6	7	7	6	7	7
4	4	6	1	2	5	4	7	5	4

Calculate:
(a) the total number of peas
(b) the average number of peas per pod
(c) the percentage of pods with less than five peas
(d) the probability that a pod picked at random would contain less than five peas.

(2) A passage of a book was chosen at random and the number of letters in each word counted as follows.

2	3	2	6	2
8	5	4	4	6
6	2	6	6	2
6	7	3	3	4
2	7	2	3	5

Calculate:
(a) the percentage of words with more than five letters

(b) the probability that a word chosen at random from this passage will have six letters or more

(c) the average number of letters per word

(d) the percentage of words which have less than the average number of letters.

(3) A bag contains 6 white balls, 4 black balls, 5 green balls and 5 red balls. If a ball is chosen at random, what is the probability of picking

(a) a white ball

(b) a red ball

(c) a ball which is not green?

(4) A light engineering firm has two factories. The Sheffield factory employs 23 people and the average weekly wage is £35.62. The Leeds factory employs 35 people and the average weekly wage is £37.71. Calculate:

(a) the total earnings at the Sheffield factory

(b) the total earnings at the Leeds factory

(c) the mean average weekly wage for all the firm's workers, correct to the nearest penny.

(5) An unbiased 20-sided die is numbered from 1 to 20. If the die is rolled, calculate the probability of scoring

(a) an odd number

(b) a number with a 1 in it

(c) 15

(d) less than 15.

50 Constructions

Perpendiculars and Bisectors

The figures given below should remind you of the constructions which are to be done using only a straight edge and a pair of compasses.

Perpendicular bisector of AB

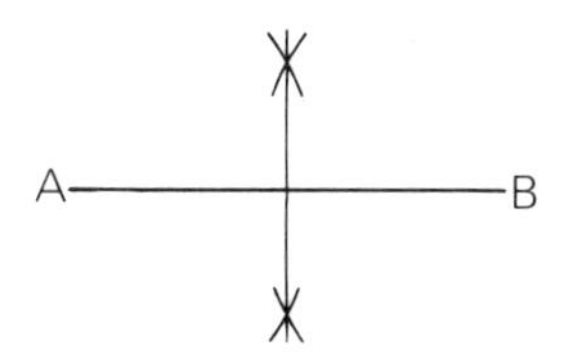

1. Draw arcs centre A.
2. Draw arcs centre B of the same radius.
3. Join intersections of arcs.

Perpendicular from P to AB

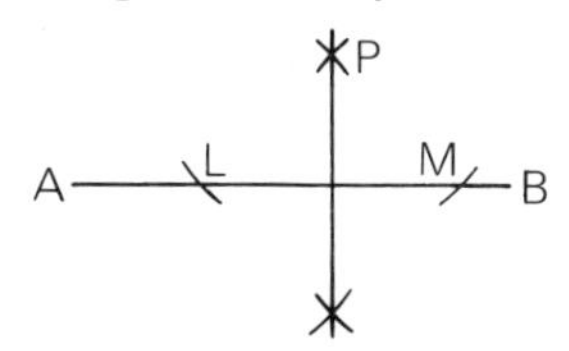

1. Draw arcs with centre P to cut AB at L and M.
2. Construct ⊥ bisector of LM.

Perpendicular to AB at P

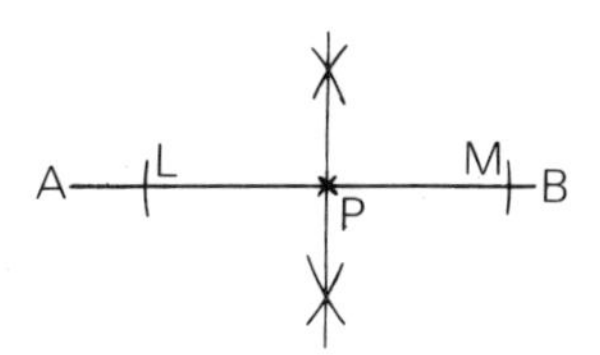

1. Draw arcs with centre P to cut AB at L and M.
2. Construct ⊥ bisector of LM.

Bisector of $\angle AOB$

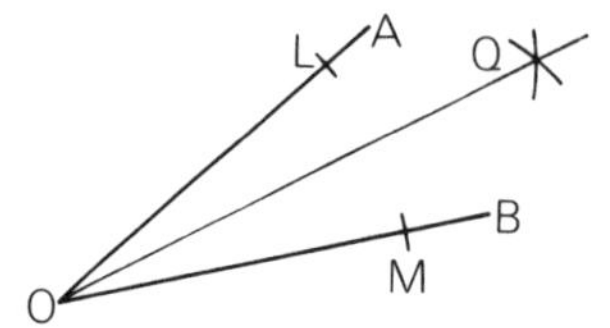

1. Draw arcs centre O, cutting OA and OB at L and M.
2. Draw arcs centre L and M to intersect at Q.
3. Join OQ.

Special Angles

A 45° angle is constructed by bisecting a right angle. A 60° angle is constructed as part of an equilateral triangle. A 30° angle is constructed by bisecting a 60° angle.

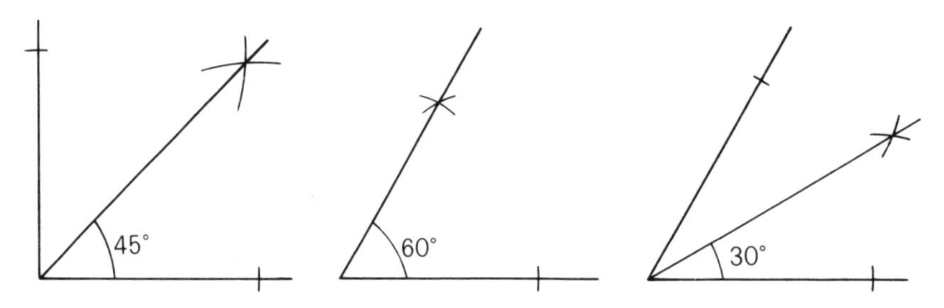

Triangles and Circles

Methods for drawing triangles are given in chapter 16. The circles connected with triangles are drawn as shown below.

Circumcircle

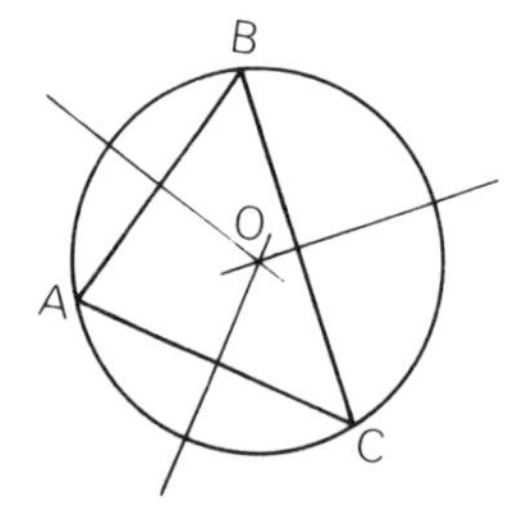

The circumcircle passes through the vertices of the triangle. The circumcentre is found by drawing the perpendicular bisectors of the sides of the triangle.

In-circle

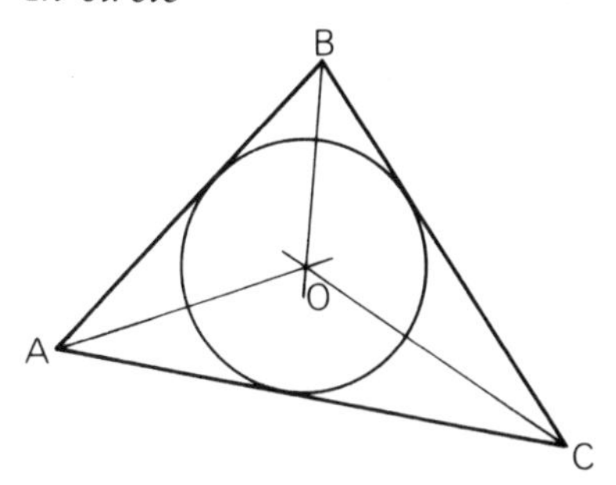

The in-circle touches the sides of the triangle. The in-centre is found by bisecting the angles of the triangle.

E-circle

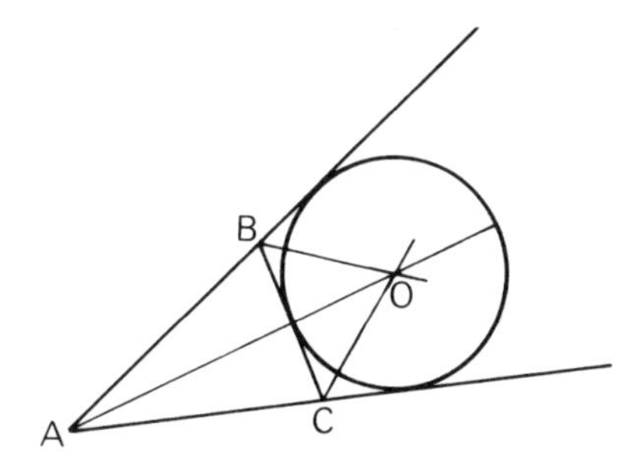

The E-circle is outside the triangle. There are three E-circles to every triangle. The one drawn here is the E-circle opposite A.

Dividing a Line

Suppose we have to divide a line AB into 5 equal parts. This can be done as shown below.

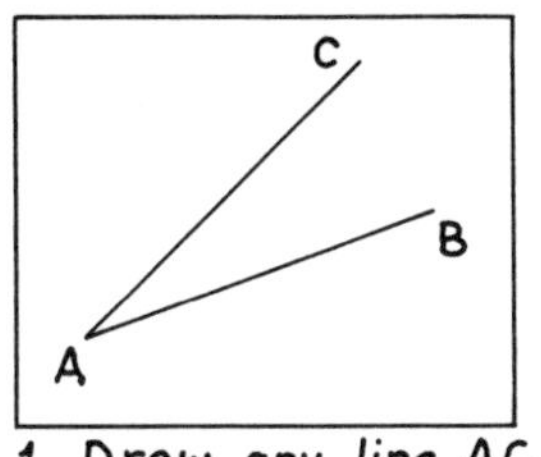

1. Draw any line AC, through A.

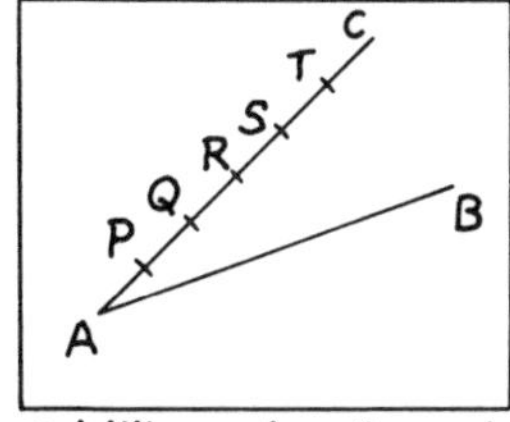

2. With centre A and any radius, draw an arc to cut AC at P. Repeat, with same radius, and centres at P, Q, R, S.

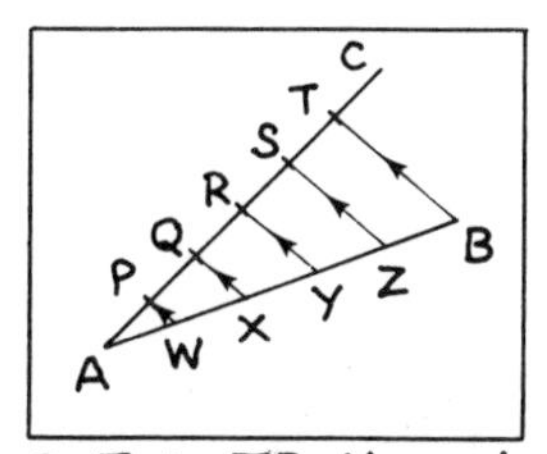

3. Join TB. Use set-square and ruler to draw PW, QX, RY, SZ parallel to TB. Then AW = WX = XY = YZ = ZB.

The triangles APW, AQX, ARY, ASZ and ATB are all similar. Can you see why? The distances AP, PQ, QR, RS and ST are *intercepts* on AC. Because they are equal, the intercepts on AB are also equal. Thus, because AT is divided into 5 equal parts AB is also divided into 5 equal parts. The same method can be used to divide a line into 6, 7 or any other number of parts.

Exercise 50a

(1) Draw a line 5 cm long and construct the perpendicular bisector.
(2) Draw a line 7 cm long and construct the perpendicular bisector.
(3) Draw a line AB 10 cm long and mark a point P about 4 cm from AB. Construct a line through P which is perpendicular to AB.
(4) Draw a line AB 10 cm long and mark a point P on AB, 4 cm from A. Construct a line through P which is perpendicular to AB.
(5) Draw a line AB, 7 cm long and mark a point P on AB, 3 cm from A. Draw a line through P which is perpendicular to AB.
(6) Draw an angle which is roughly 60° and construct the bisector of the angle.
(7) Draw an obtuse angle and, using the same construction as before, bisect the angle.
(8) Draw a reflex angle and, using the same construction as before, bisect the angle.
(9) Draw a triangle ABC in which AB = 6 cm, BC = 7 cm and AC = 10 cm. Through B, draw a line perpendicular to AC, to cut BC at P. Measure BP, to the nearest mm.
(10) Draw triangle PQR in which PQ = QR = PR = 7 cm. Construct all three altitudes.
(11) Draw triangle ABC in which AB = 5 cm, BC = 7 cm and AC = 10 cm. Construct all three altitudes.
(12) Draw triangle KLM in which KL = 6.3 cm, LM = 5.5 cm and KM = 7.9 cm. Construct the perpendicular bisector of each side and draw the circumcircle and measure its radius.
(13) Repeat question 12 for △ABC in which AB = 10.5 cm, BC = 7.5 cm and AC = 6 cm.
(14) Repeat question 12 for △PQR in which PQ = 6 cm, QR = 8 cm and PR = 10 cm.
(15) Draw an equilateral triangle of side 8 cm and construct the angle bisectors. Draw the in-circle and measure its radius.
(16) Repeat question 15 for △ABC in which AB = 9 cm, BC = 7 cm and AC = 10.5 cm.
(17) Draw an equilateral triangle of side 5 cm and construct one E-circle. Measure its radius.

(18) Draw △PQR in which PQ = 6 cm, QR = 8 cm and PR = 10 cm. Construct the E-circle opposite Q.
(19) Draw △XYZ in which XY = 6 cm, XZ = 3 cm and YZ = 4 cm. Construct the E-circle opposite X.
(20) How far apart are the lines of your exercise book? Can you draw a line across the lines of your exercise book so that the intercepts are (i) 1 cm long (ii) 1.5 cm long (iii) 2 cm long?
(21) Draw a line AB = 10 cm. Use the construction shown on page 241 to divide the line into seven equal parts.
(22) Draw a line OP = 7 cm and without measuring, divide the line into five equal parts.
(23) Draw a line PQ = 12 cm. Use the construction shown on page 241 to divide PQ into eleven equal parts.
(24) Draw a line QT = 8 cm and divide it into three equal parts without measuring.
(25) Draw a line 11 cm long and using the construction shown on page 241, find the mid-point of the line.

Exercise 50b

(1) Draw △ABC in which AB = BC = 4.5 cm and AC = 8 cm.
(a) Bisect ∠ABC and mark a point D on your bisector so that BD = 9 cm.
(b) Join AD and CD. Name the quadrilateral you have drawn.
(c) Measure AD correct to the nearest mm.
(2) Draw a line AB 10.5 cm long and construct the perpendicular bisector.
(a) Mark points P and Q on the bisector such that P and Q are both 4.3 cm from the mid-point of AB.
(b) Join AP, PB, BQ and QA. Name the quadrilateral you have drawn.
(c) Measure AP correct to the nearest mm.
(d) Calculate the area of the figure correct to 3 significant figures.
(3) Draw △XYZ in which XY = 6 cm, YZ = 7 cm and XZ = 8 cm.
(a) Construct the altitude through X, to cut YZ at P. Measure XP.
(b) Construct the altitude through Y, to cut XZ at Q.
(c) Name one pair of similar triangles in your figure.
(d) Calculate the area of △XYZ correct to 2 significant figures.
(4) Draw △ABC in which AB = 7 cm, BC = 6 cm and AC = 8 cm.
(a) Construct the perpendicular bisector of each side and draw the circumcircle. Measure its radius.
(b) Construct the E-circle opposite A. Measure its radius.
(c) Measure the distance between the circumcentre and the E-centre.

Section I

51 Quadratic Equations

Solving by Factors

If the quadratic function $ax^2 + bx + c$ can be factorised, the quadratic equation $ax^2 + bx + c = 0$ can be solved. Usually, there are two possible values of x which satisfy a quadratic equation and these are called the *roots* of the equation.

Example: Solve the equation $2x^2 - x - 10 = 0$.

If $\quad 2x^2 - x - 10 = 0$
then $\quad (2x - 5)(x + 2) = 0.$

$\therefore$ either $\quad (2x - 5) = 0$
or $\quad (x + 2) = 0.$

If
$$2x - 5 = 0$$
$$2x = 5$$
$$x = 2\tfrac{1}{2}.$$

If
$$x + 2 = 0$$
$$x = -2.$$

If the left-hand side of the equation is a perfect square, the two roots have the same value.

Example: Solve the equation $x^2 - 6x + 9 = 0$.

If $\quad x^2 - 6x + 9 = 0$
then $\quad (x - 3)(x - 3) = 0.$

$\therefore$
$$x - 3 = 0$$
$$x = 3.$$

Solving by Graphs

Some quadratic equations can be solved by drawing a graph, as the following example shows.

Example: Solve graphically the equation $10x^2 - x - 3 = 0$.

First, draw the graph of $y = 10x^2 - x - 3$.

x	-3	-2	-1	0	1	2	3
$10x^2$	90	40	10	0	10	40	90
$-x$	3	2	1	0	-1	-2	-3
-3	-3	-3	-3	-3	-3	-3	-3
y	90	39	8	-3	6	35	84

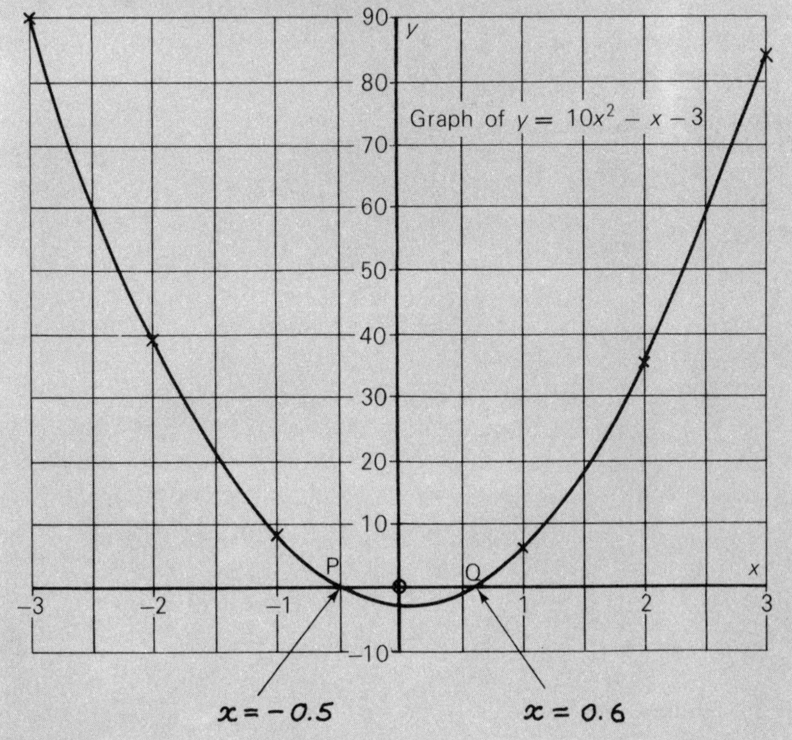

At all points on the graph, $y = 10x^2 - x - 3$.
At all points on the x-axis, $y = 0$.
Therefore, at P and Q, $10x^2 - x - 3 = 0$.
The x co-ordinates of P and Q satisfy the equation $10x^2 - x - 3 = 0$ so that the roots are $x = -0.5$ and $x = 0.6$.

Exercise 51a

Solve by factors.

(1) $x^2 - x - 2 = 0$
(2) $x^2 + 4x + 3 = 0$
(3) $x^2 - 2x + 1 = 0$
(4) $x^2 + 4x - 5 = 0$
(5) $x^2 - 6x - 7 = 0$
(6) $x^2 - x - 6 = 0$
(7) $x^2 - 7x + 10 = 0$
(8) $x^2 - 4x - 21 = 0$
(9) $x^2 - 2x - 15 = 0$
(10) $x^2 + 2x - 8 = 0$
(11) $x^2 + x - 6 = 0$
(12) $x^2 - 5x + 6 = 0$
(13) $x^2 + 5x + 6 = 0$
(14) $x^2 - 5x - 6 = 0$
(15) $x^2 - 9x + 8 = 0$
(16) $x^2 + 4x + 4 = 0$
(17) $3x^2 - 5x - 2 = 0$
(18) $x^2 - 6x + 9 = 0$
(19) $5x^2 - 14x - 3 = 0$
(20) $3x^2 + 8x - 3 = 0$
(21) $2x^2 - x - 1 = 0$
(22) $2x^2 + 5x - 3 = 0$
(23) $3x^2 - x - 2 = 0$
(24) $7x^2 - 13x - 2 = 0$
(25) $11x^2 + 9x - 2 = 0$
(26) $2x^2 - x - 3 = 0$
(27) $3x^2 - 5x + 2 = 0$
(28) $2x^2 - x - 6 = 0$
(29) $2x^2 + 7x - 15 = 0$
(30) $3x^2 + 4x - 4 = 0$

Solve the following equations graphically.

(31) $x^2 - 3x + 2 = 0$
(32) $2x^2 + x - 1 = 0$
(33) $2x^2 - 3x - 2 = 0$
(34) $5x^2 + 4x - 1 = 0$
(35) $2x^2 + x - 6 = 0$
(36) $4x^2 - 1 = 0$
(37) $2x^2 - x - 10 = 0$
(38) $2x^2 + x - 3 = 0$
(39) $4x^2 - 4x + 1 = 0$
(40) $x^2 - x - 2 = 0$

Exercise 51b

(1) Solve the following equations.
(a) $2x - 3 = 0$.
(b) $x^2 - 8x + 7 = 0$.
(c) $3x^2 + 13x - 10 = 0$.

(2) The length of a rectangle is $(x + 1)$ cm and the width is $(x - 2)$ cm.

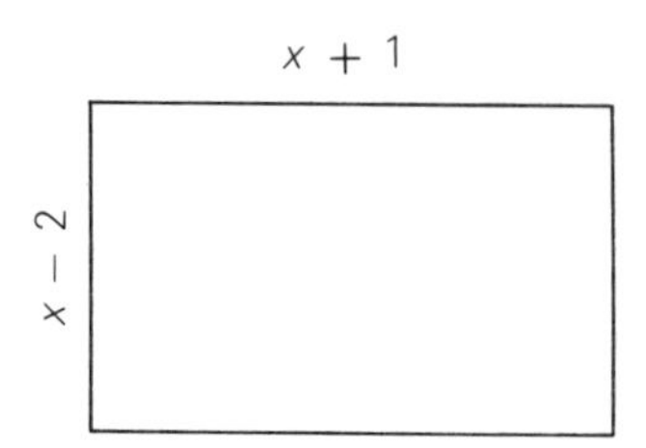

(a) Make an expression for the area of the rectangle.

(b) If the area of the rectangle is 4 cm^2 make up a quadratic equation and solve it to show that there is only one possible value for x.

(c) Calculate the length of the diagonal to the nearest mm.

(3) ABC is a right-angled triangle with sides as shown.

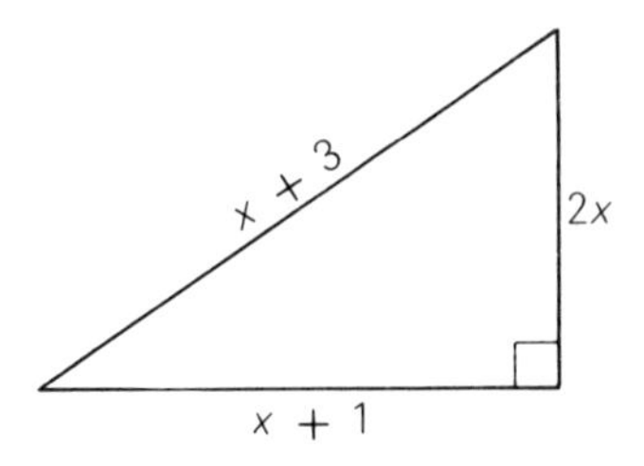

(a) Find an expression for the perimeter of the triangle in its simplest form.

(b) Write down an expression for the area of the triangle in descending powers of x.

(c) Make a quadratic equation in x by using the Pythagoras rule.

(d) Solve your equation and write down the radius of the circum-circle of the triangle.

(4) Draw the graph of $y = x^2 - 4x + 3$ from $x = -2$ to $x = +5$.

(a) Write down the co-ordinates of the turning point and say whether it is a maximum or minimum point.

(b) Find the gradient of the curve at $x = 0$.

(c) Use your graph to solve the equation $x^2 - 4x + 3 = 0$.

(5) Draw the graph of $y = 2x - x^2$ from $x = -3$ to $x = +3$.

(a) Write down the co-ordinates of the turning point and say whether it is a maximum or minimum point.

(b) Find the gradient of the curve at $x = 2$.

(c) Use your graph to solve the equation $2x - x^2 = 0$.

52 Right-Angled Triangles

Heights and Distances

A man stands on the beach and looks up to his friend who is on top of a cliff.

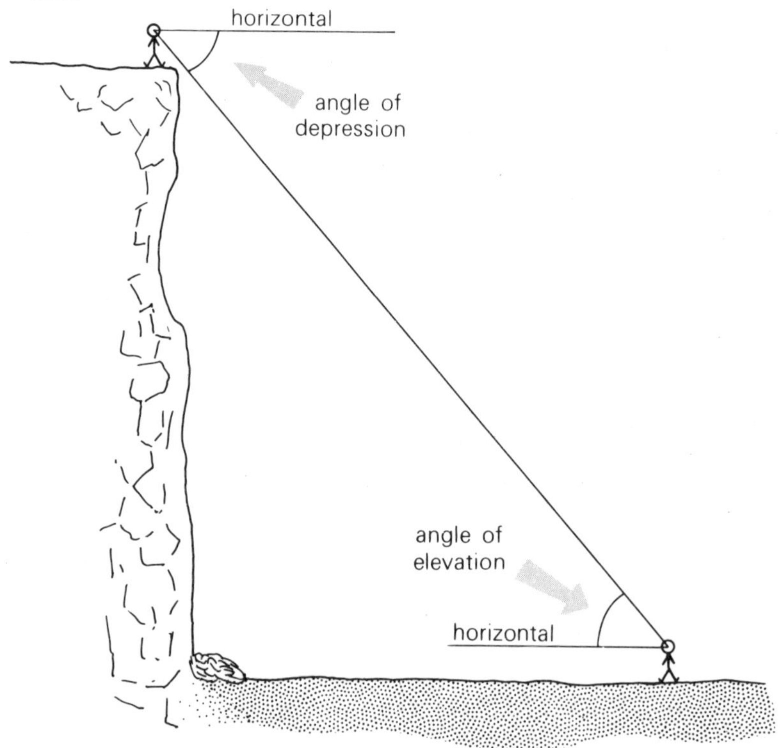

For the man looking up, the angle between his line of sight and the horizontal is called the *angle of elevation*. For the man looking down, the angle between his line of sight and the horizontal is called the *angle of depression*. These two angles are alternate angles and therefore equal. Angles of elevation and depression can be used for calculating heights.

Example: A man lying on the edge of a cliff looks out to a boat at sea. If the angle of depression of his line of sight is 13° and the boat is 197 m out to sea, calculate the height of the cliff to the nearest metre.

$$\tan 13° = \frac{h}{197}$$

$\therefore \quad h = 197 \tan 13°$

$= 197 \times 0.2309$ (from tables)

$= 45.48$ (by logs)

$= 45$ m (correct to the nearest metre).

No	Log
0.2309	$\bar{1}.3634$
197	2.2945
45.48	1.6579

Solution of Triangles

If some facts about a triangle are known, it may be possible to find the others. If we do this, we *solve* the triangle.

Example: In $\triangle ABC$, AB = 12.7 cm, AC = 5.9 cm and $\angle C = 90°$. Solve the triangle completely.

$$\sin B = \frac{5.9}{12.7}$$

$= 0.4646$ (by logs).

$\therefore \angle B = 27° 41'$ (from sine tables).

No	Log
5.9	0.7709
12.7	1.1038
0.4646	$\bar{1}.6671$

$\angle A = 90° - \angle B$

$= 90° - 27° 41'$

$= 62° 19'$.

$BC^2 = AB^2 - AC^2$ (Pythagoras)

$= 161.3 - 34.81$ (by logs or from tables of squares)

$= 126.5$

$\therefore BC = 11.24$ (from square root tables)

$= 11.2$ (correct to 3 significant figures).

In the above example $\angle B$ has been found before $\angle A$ by finding sin B. The triangle could have been solved equally well by finding cos A in which case $\angle A$ would have been found before $\angle B$.

Special Angles

ABCD is a square. The diagonal BD bisects the angles at the corners, so that $\triangle ABD$ is a right-angled triangle, with $\angle ABD = \angle ADB = 45°$. If we take AB = AD to be 1 unit then

$$BD^2 = 1^2 + 1^2 \text{ (Pythagoras)}$$
$$= 2$$

and $BD = \sqrt{2}$.

From this, we can see that $\sin 45° = \frac{1}{\sqrt{2}}$, $\cos 45° = \frac{1}{\sqrt{2}}$, and $\tan 45° = 1$.

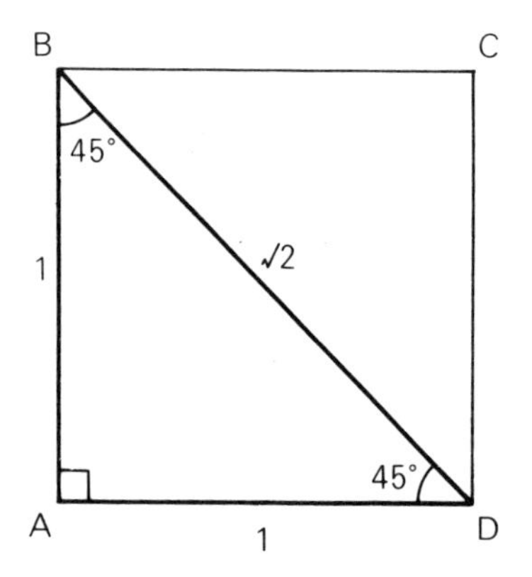

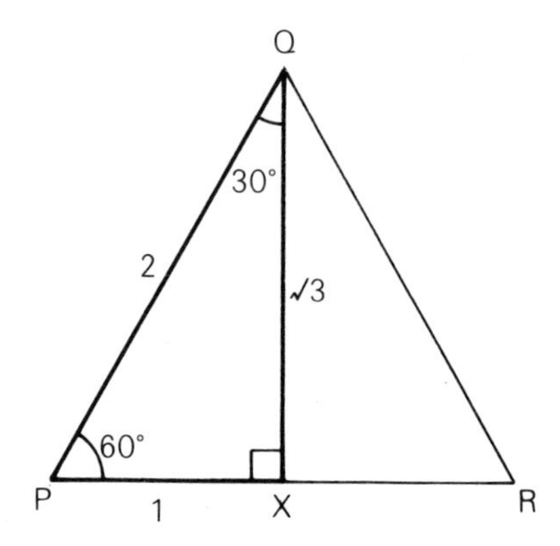

$\triangle PQR$ is an equilateral triangle. X is the mid-point of PR, so that $QX \perp PR$. This means that PQX is a right-angled triangle with $\angle P = 60°$, so that $\angle PQX$ is 30°. If we take PX to be 1 unit, then PQ = 2 units and

$$QX^2 = PQ^2 - PX^2$$
$$= 4 - 1$$
$$QX^2 = 3$$
$$QX = \sqrt{3}.$$

From this we can see that $\sin 60° = \frac{\sqrt{3}}{2}$, $\cos 60° = \frac{1}{2}$ and $\tan 60° = \sqrt{3}$. Also $\sin 30° = \frac{1}{2}$, $\cos 30° = \frac{\sqrt{3}}{2}$ and $\tan 30° = \frac{1}{\sqrt{3}}$. These results are summarised in the following table.

angle	sine	cosine	tangent
30°	$\frac{1}{2}$	$\frac{\sqrt{3}}{2}$	$\frac{1}{\sqrt{3}}$
45°	$\frac{1}{\sqrt{2}}$	$\frac{1}{\sqrt{2}}$	1
60°	$\frac{\sqrt{3}}{2}$	$\frac{1}{2}$	$\sqrt{3}$

If you look in your square root tables you will find that $\sqrt{2} = 1.414$, so that $\sin 45° = \frac{1}{\sqrt{2}} = \frac{1}{1.414} = 0.707$ (by logs). You can check this with your sine tables, but of course, it is only accurate to 3 significant figures. If we write $\sin 45° = \frac{1}{\sqrt{2}}$, we know exactly what we mean. If we write $\sin 45° = 0.707$ this is only an approximation. For this reason, the sine, cosine and tangent of 30°, 45° and 60° are often used in square root form.

Example: If the side of an equilateral triangle is 6 cm long, calculate (i) the altitude (ii) the area of the triangle.

(i) The altitude of the triangle is its height, BX

$$\sin 60° = \frac{h}{6}$$

$$\therefore \frac{\sqrt{3}}{2} = \frac{h}{6}$$

$$h = \frac{6\sqrt{3}}{2}$$

$$= 3\sqrt{3} \text{ cm.}$$

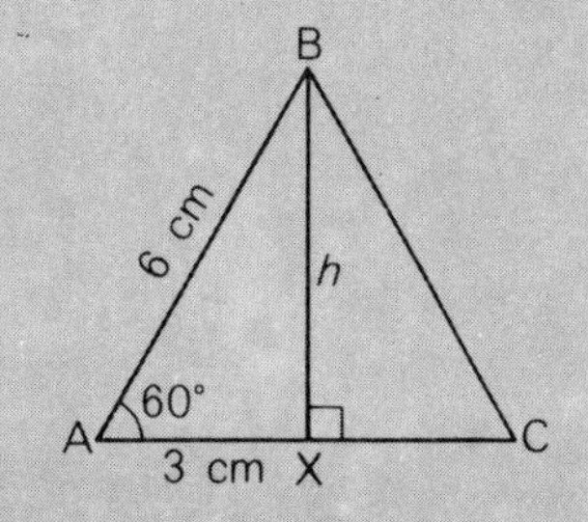

(ii) Area of $\triangle$ABC is $\frac{1}{2}$ base $\times$ height

$$= \tfrac{1}{2}\,\text{AC}\cdot\text{BX}$$

$$= \tfrac{1}{2} \times 6 \times 3\sqrt{3}$$

$$= 9\sqrt{3} \text{ cm}^2.$$

If the length of the side of a square is x units, the length of the diagonal is $x\sqrt{2}$ units. If the length of the side of an equilateral triangle is x units, the altitude of the triangle is $\frac{x\sqrt{3}}{2}$ units.

Exercise 52a

Each of the following represents a triangle ABC with $\angle C = 90°$. Solve the triangles completely.

(1) $a = 4.7$ cm, $b = 7.3$ cm
(2) $a = 11.9$ cm, $b = 2.4$ cm
(3) $a = 5.5$ cm, $c = 7.5$ cm
(4) $a = 3.7$ cm, $c = 8.1$ cm
(5) $b = 23.5$ cm, $c = 35.2$ cm
(6) $\angle A = 25°$, $b = 7.6$ cm
(7) $\angle A = 40°$, $b = 8.5$ cm
(8) $\angle A = 62°$, $c = 7.3$ cm
(9) $\angle B = 82°$, $c = 6.1$ cm
(10) $\angle B = 43°$, $c = 11.2$ cm.

(11) From a point at ground level, the angle of elevation of the top of a building 30 m away is seen to be 20°. Calculate the height of the building to the nearest metre.

(12) From a window 5 m above ground level, the angle of elevation of the top of a tree is seen to be 22°. If the tree is 33.5 m from the window, find the height of the tree.

(13) From an observation post at ground level, a man sees an aeroplane fly over the control tower of an airport at an angle of elevation of 53°. If the height of the aircraft is 1 200 m, calculate the distance of the observation post from the control tower to the nearest metre.

(14) From the top of a radio mast the angle of depression of a point at ground level 275 m from the mast, is found to be 13°. Calculate the height of the mast to the nearest metre.

(15) A flagpole 15.5 m tall throws a shadow which is 12.7 m long. Calculate the angle of elevation of the sun.

(16) Calculate the length of shadow cast by a vertical post 3.7 m tall when the angle of elevation of the sun is 29°.

(17) A rescue helicopter sights a dinghy at an angle of depression of 27°. If the dinghy is 95 m from the point directly under the helicopter calculate the height of the helicopter to the nearest metre.

(18) The base of an isosceles triangle is 3 cm long and its altitude 4 cm. Calculate the angles of the triangle.

(19) An aeroplane flies 40 km due east from High Prairie, then 100 km north to Beaver Creek. Calculate the distance and bearing of Beaver Creek from High Prairie.

(20) A plane flies 60 km on a course of 217° and then 25 km on a course of 307°. Calculate the distance and bearing of the aircraft from its starting position.

(21) A ladder 5 m long rests against a wall at an angle of 50° to the horizontal ground. How far up the wall is the top of the ladder?

(22) A guy rope is fixed to the top of a tent pole and the other end is fixed to the ground so that the rope, which is straight, makes an angle of 35° to the horizontal ground. If the rope is 5.5 m long, calculate the height of the pole to the nearest cm.

(23) The length of a sloping roof from ridge to gutter is 6.3 m. If the roof makes an angle of 32° to the horizontal, calculate the height of the ridge above the gutter.
(24) The diagonal of a square is 12.7 cm long. Calculate the length of the side of the square to the nearest mm.
(25) The sides of an isosceles triangle are 8 cm, 9.5 cm and 9.5 cm. Calculate all the angles of the triangle.
(26) An aeroplane flies 75 km on a bearing of N 53° E. How far east has it travelled?
(27) A plane flies 127 km on a bearing of 245°. How far west has it travelled?
(28) A plane flies 97 km on a bearing of 305°. How far north has the plane travelled?
(29) Calculate the diagonal of a square if its side is 43 cm long.
(30) Calculate the altitude of an equilateral triangle of side 16 cm.

Exercise 52b

(1) In △ABC, calculate:

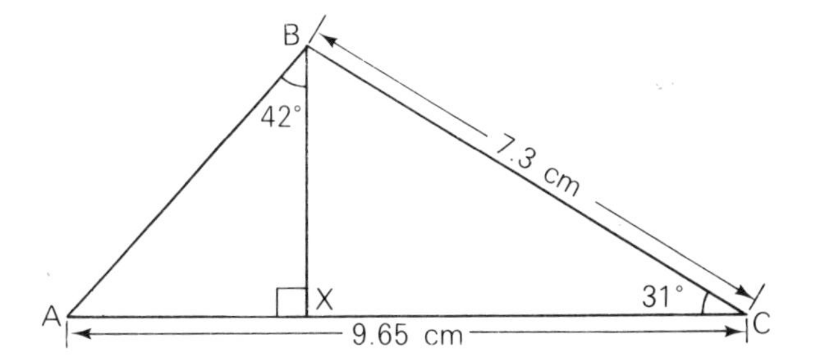

(a) ∠ABC and ∠BAC
(b) BX
(c) the area of △ABC
(d) the side of the square whose area is equal to the area of △ABC.

(2) A plane flies 99 km from Yeadon airport in a direction 110° and then 173 km due south to Luton. Find:
(a) how much farther east Luton is than Yeadon
(b) how much farther south Luton is than Yeadon
(c) the distance from Yeadon to Luton
(d) the bearing of Yeadon from Luton.

Give all distances correct to the nearest km.

(3) The diagonals of a square ABCD intersect at X. If AB $= 4\sqrt{2}$ cm, calculate:
(a) the length of AC

(b) the area of ABCD
(c) the area of the circle on AB as diameter.
Give exact answers to (a) and (b) and give your answer to (c) as a multiple of π.

(4) A yachtsman notices a coastguard watching him from the top of a cliff. If the angle of depression of the coastguard's line of sight is 39° and if the distance between the two men is 87 m, find:

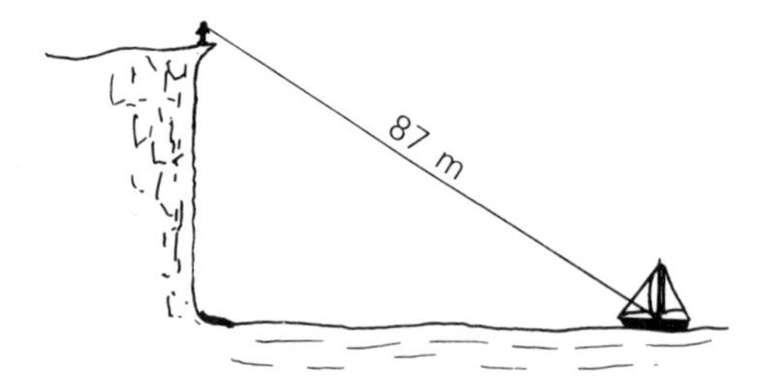

(a) the angle of elevation of the yachtsman's line of sight
(b) the height of the cliff to the nearest metre
(c) the distance from the boat to the cliff, correct to the nearest metre.

(5) ABC is an isosceles triangle and X is the mid-point of AC. If AC = 9.6 cm and BX = 4.5 cm, calculate:
(a) $\angle$ACB
(b) $\angle$ABC
(c) the length of BC
(d) the area of $\triangle$ABC.

53 Circles

Chords

The perpendicular bisector of a chord passes through the centre of the circle. In (a), AX = XB and $\angle$OXB = 90°.

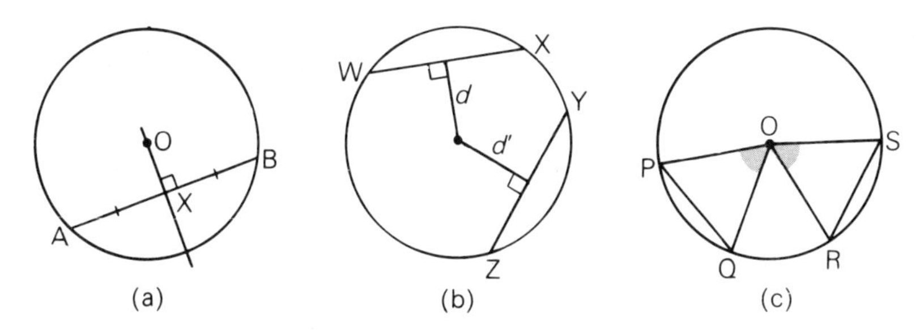

(a) (b) (c)

Equal chords are at equal distances from the centre of the circle. Chords which are at equal distances from the centre of the circle are equal in length. In (b), WX = YZ and $d = d'$.

Equal chords *subtend* equal angles at the centre of the circle. In (c), PQ = RS and the grey angles are equal.

Angles

Angles in the same segment are equal. In (d) all the grey angles are equal.

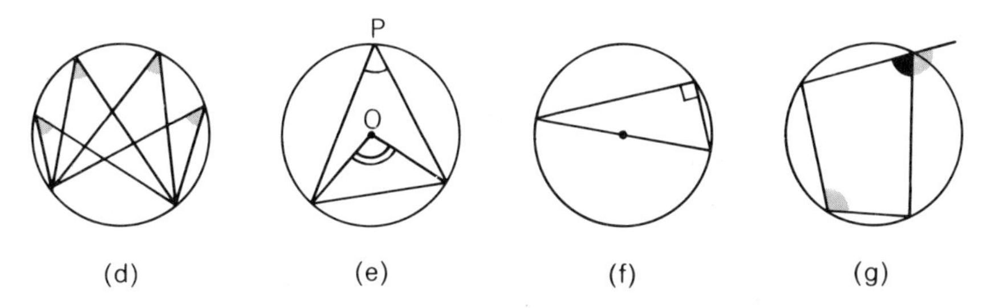

(d) (e) (f) (g)

The angle at the centre of the circle is twice the angle at the circumference subtended by the same chord. In (e), the angle at O is twice the angle at P. In particular if the chord is a diameter as in (f), the angle in the semi-circle is a right angle.

A quadrilateral which has its vertices on the circumference of a circle is called a *cyclic quadrilateral*. Each pair of opposite angles is *supplementary*. That is, a pair of opposite angles adds up to 180°. In (g) the black angle added to a grey angle makes a total of 180°.

In (g) one grey angle is outside the quadrilateral and is an *exterior* angle. The other grey angle is the opposite *interior* angle. In any cyclic quadrilateral an exterior angle is equal to the opposite interior angle.

Intersecting Chords

In (h), AB and CD are chords intersecting at X. In (i), the chords intersect outside the circle.

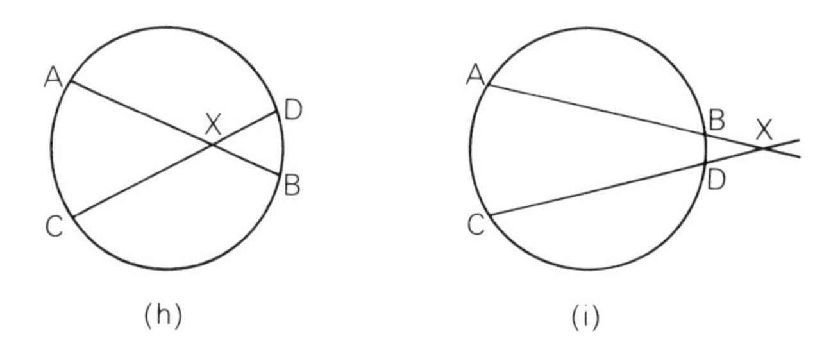

(h) (i)

In both cases $AX \cdot XB = CX \cdot XD$.

Tangents

A tangent touches the circle at a *point of contact*. If a radius is drawn through the point of contact it is perpendicular to the tangent. From any point outside the circle, two tangents can be drawn to the circle. These tangents are equal in length.

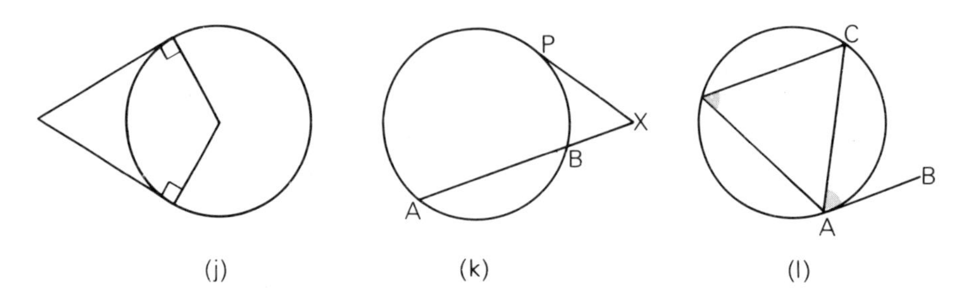

(j) (k) (l)

In (k), AB is a chord which intersects the tangent at P outside the circle. In this case, $AX \cdot BX = PX^2$.

In (l), AC is a chord drawn through the point of contact of the tangent AB. The angle between the tangent and the chord is equal to the angle in the alternate segment.

Exercise 53a

(1) AB is a chord of a circle centre O, radius 7.5 cm. Calculate the length of the chord if it is 3.6 cm from O.
(2) Calculate $\angle OAB$ in question 1.
(3) Calculate the angle subtended at the centre of a circle radius 5 cm by a chord 8 cm long.
(4) ABCD is a cyclic quadrilateral in a circle centre O. If $\angle ABD = 32°$, calculate $\angle ACD$.
(5) Calculate $\angle AOD$ in question 4.
(6) PQ is a diameter of a circle centre O. R is a point on the circumference such that $\angle POR = 48°$, find $\angle PQR$.
(7) Calculate $\angle QPR$ in question 6.
(8) ABCD is a cyclic quadrilateral in a circle centre O. If AOC is a straight line, write down the size of $\angle B$.
(9) In a cyclic quadrilateral PQRS, if $\angle Q = 49$, what size is $\angle S$?
(10) In a cyclic quadrilateral KLMN, if $\angle L = 130°$, what size is $\angle N$?
(11) ABCD is a cyclic quadrilateral and the diagonals intersect at X. If AX = 3 cm, XC = 4 cm and BX = 2 cm, calculate XD.
(12) Two chords AB and CD of a circle intersect at a point X outside the circle. If AX = 12 cm, BX = 3 cm and CX = 9 cm, calculate the length of DX.
(13) Two chords PQ and RS of a circle intersect at a point T outside the circle. If PT = 8 cm, QT = 5 cm and ST = 2.5 cm calculate the length of RT.
(14) Two chords KL and MN of a circle intersect at X outside the circle. If KL = 2 cm, LX = 6 cm and NX = 4 cm, calculate MN.
(15) O is the circumcentre of $\triangle ABC$. If $\angle B = 73°$, calculate $\angle AOC$.

Questions 16–25 refer to the figure at the top of page 258 in which O is the centre of the circle and PT is a tangent.

(16) What is the size of $\angle SRQ$?
(17) If $\angle TPR = 60°$, what size is $\angle PSR$?
(18) If $\angle SPR = 45°$, what size is $\angle SQR$?
(19) If $\angle SPR = 53°$, what size is $\angle SOR$?
(20) If OS = 7 cm and SR = 10 cm, use tables to calculate $\angle OSR$.
(21) If PX = 5 cm, XR = 4 cm and SX = 8 cm, calculate XQ.

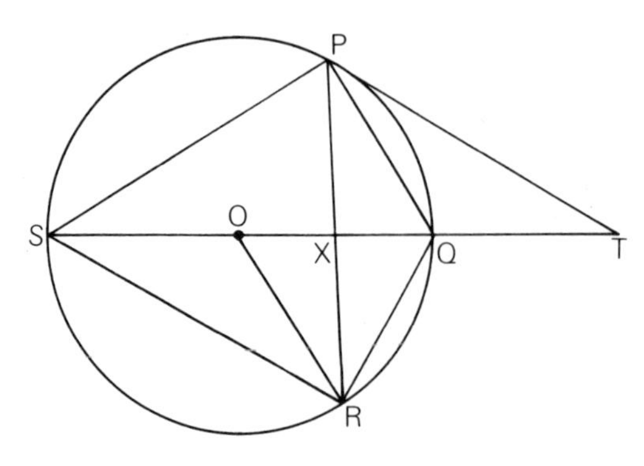

(22) If $\angle SPQ = 78°$, calculate $\angle SRQ$.
(23) If $\angle PQS = 42°$, calculate $\angle PRS$.
(24) If $\angle PXS = 81°$, calculate $\angle QXR$.
(25) If $QT = 4$ cm and $ST = 16$ cm, calculate PT.

Exercise 53b

(1) O is the circumcentre of $\triangle ABC$ and X is the mid-point of AC. If $AO = 7.5$ cm and $AC = 6.9$ cm, calculate: (a) $\angle AOX$ (b) $\angle AOC$ and $\angle ABC$ (c) $\angle ACO$ (d) OX.
(2) In the figure, ABCD is a cyclic quadrilateral and O is the centre of the circle. Calculate:

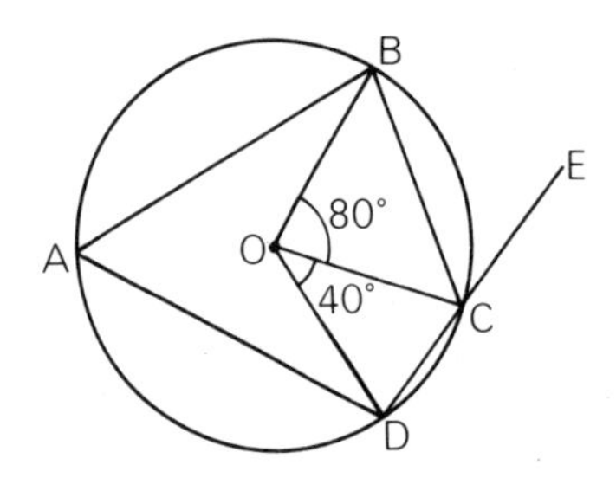

(a) $\angle DAB$ (b) $\angle BCE$ (c) $\angle OCD$ (d) $\angle OBC$.
(3) In the figure PT is a tangent to a circle centre O. If $PT = 15$ cm and $OT = 17$ cm, calculate:

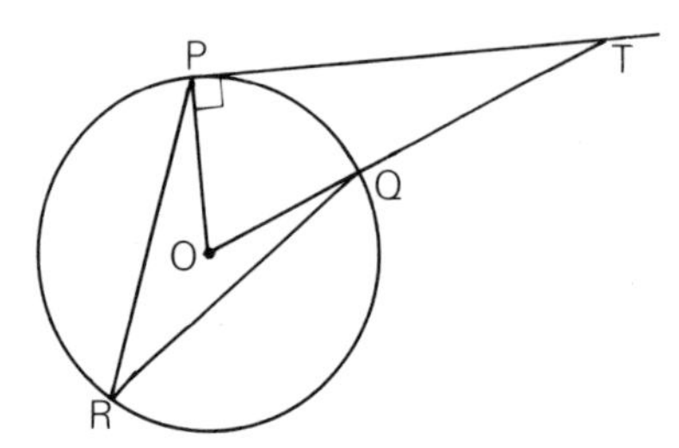

(a) OP (b) QT (c) $\angle PTQ$ (d) $\angle PRQ$.

(4) If BD is a tangent to the circle centre O and $\angle BOC = 120°$, calculate:

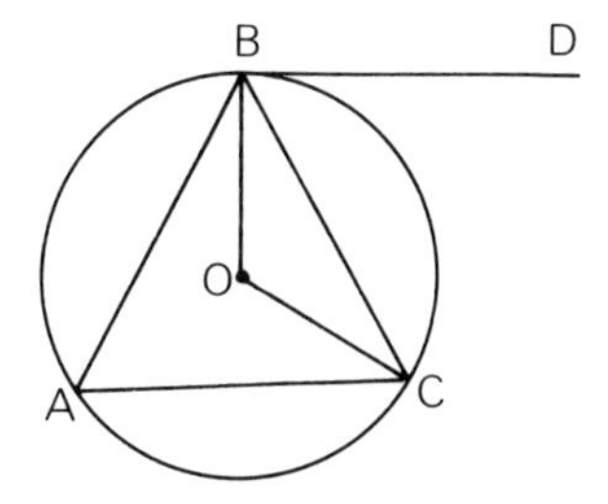

(a) $\angle BAC$ (b) $\angle DBC$ (c) $\angle BCO$.

(5) If, in the figure, BX = 8 cm, XD = 3 cm, AX = 5 cm, $\angle CBD = 25°$ and $\angle BCD = 114°$, calculate:

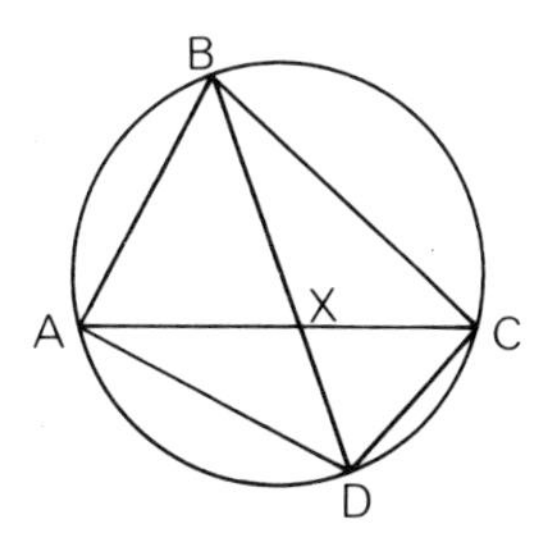

(a) XC (b) $\angle CAD$ (c) $\angle BAD$ (d) $\angle BDC$.

54 Travel

Speed

On any journey the speed varies. The *average speed* for the journey is

$$\text{average speed} = \frac{\text{distance travelled}}{\text{time taken}}.$$

If the distance is measured in cm and time in seconds, the speed is in cm/s. If the distance is measured in metres and the time in seconds, the speed is in m/s. If the distance is measured in km and the time in hours, the distance is in km/h.

Because $\text{speed} = \frac{\text{distance}}{\text{time}}$,

we can also say that $\text{distance} = \text{speed} \times \text{time}$

and $\text{time} = \frac{\text{distance}}{\text{speed}}$.

Example: A car travels 420 km at an average speed of 50 km/h. Find how long the journey takes.

$$\text{Time taken is } \frac{\text{distance}}{\text{speed}}$$

$$= \frac{420}{50}$$

$$= 8.4 \text{ hours}$$

$$= 8 \text{ hours } 24 \text{ mins } (0.1 \text{ h} = 6 \text{ mins}).$$

Distance-Time Graph

If we consider only average or constant speeds for a journey, the graph showing distances and times will consist of straight lines.

Example: The Greyhound coach leaves Fort St John at 07.30 and travels 65 km along the Alaska Highway to Dawson Creek at an average speed of 65 km/h. After stopping at Dawson Creek for 30 minutes the coach continues its journey to Grande Prairie, 135 km away, at an average speed of 90 km/h. Draw the distance-time graph for the journey.

Time taken from Fort St John to Dawson Creek is $\frac{65}{65} = 1$ hour.

$\therefore$ Time of arrival at Dawson Creek is 08.30.
Departure from Dawson Creek is 09.00.
Time taken from Dawson Creek to Grande Prairie is

$$\frac{135}{90} = 1\tfrac{1}{2} \text{ hours.}$$

$\therefore$ Time of arrival at Grande Prairie is 10.30.

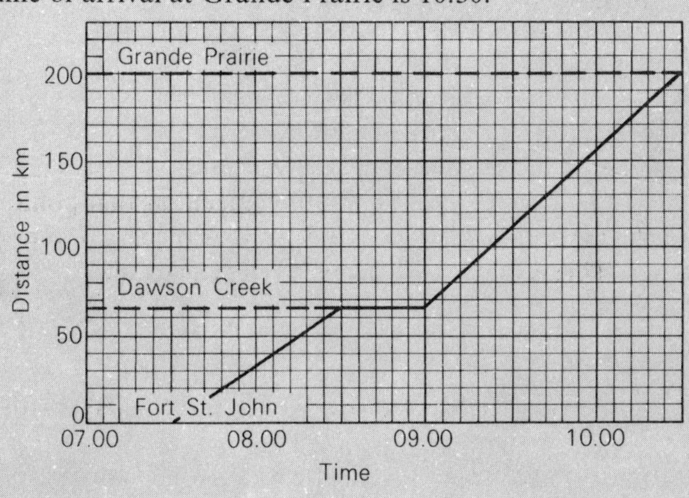

In the above example, we are only concerned with average speeds. If we want to know how fast the coach was moving at any one time, we must either read the speedometer or we must draw a more accurate graph.

The table below shows the distances travelled by the coach as it starts from rest.

Time in s	0	5	10	15	20
Distance in m	0	1	4	10	20

If the points are plotted and a smooth curve drawn through them, we have a complete picture of the movement.

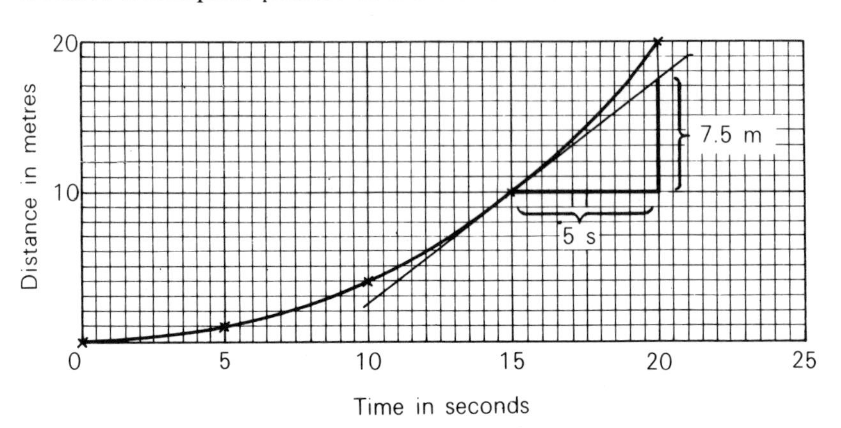

The average speed for the first 20 s is

$$\frac{\text{distance}}{\text{time}} = \frac{20}{20} = 1 \text{ m/s}.$$

This is the average speed. To get the speedometer speed after 15 s, we must find the slope of the curve at $t = 15$. We can find the slope of the curve at $t = 15$ by finding the slope of the tangent at that point.

The slope of the tangent is $7.5 \div 5 = 1.5$, so that the speed after 15 seconds is 1.5 m/s. This is the *instantaneous speed* at that time.

Exercise 54a

(1) Calculate the average speed of a car which travels 480 km in 8 hours.

(2) Calculate in m/h the average speed of a snail which travels 6 cm in 40 minutes.

(3) Calculate the average speed of a train which travels 162 km in $2\frac{1}{4}$ hours.

(4) Calculate the average speed of a swallow which flies 210 km in $1\frac{1}{2}$ hours.

(5) Calculate the distance flown by an aircraft in $3\frac{1}{2}$ hours at 450 km/h.

(6) Calculate the distance travelled by a train in $2\frac{3}{4}$ hours at 84 km/h.

(7) Calculate the distance run by a greyhound in half a minute at 12.5 m/s.
(8) Calculate the distance travelled by a car in $5\frac{1}{2}$ hours at 42 km/h.
(9) Calculate the time taken to travel 500 km at 200 km/h.
(10) Calculate the time taken to travel 500 km at 240 km/h.
(11) Calculate the time taken to drive 270 km at 60 km/h.
(12) Calculate the time taken to travel 3 000 km at 25 km/h.
(13) A man drives 45 km at 30 km/h and then 210 km at 60 km/h. Calculate his average speed for the whole journey.
(14) A boy cycles 25 km at 20 km/h and the next 30 km at 40 km/h. Calculate his average speed for the whole journey.
(15) A train travels 80 km at an average speed of 25 km/h and then returns non-stop at 100 km/h. Calculate the average speed for the round trip.
(16) A plane flies 1 000 km at 600 km/h with a tail wind and returns at 300 km/h with a head wind. Calculate the average speed for the return trip.
(17) A car travels 200 km at 60 km/h and then 20 km in 1 hour 40 minutes. Calculate the average speed for the whole journey.
(18) A train leaves Cambridge at 14.10 and travels 40 km to Hitchin in exactly 1 hour. After a 15 minute stop it then travels a further 50 km to King's Cross station in London at an average speed of 60 km/h. Draw a distance-time graph for the journey.
(19) A train leaves Hatfield at 17.00 and travels 30 km to Luton at an average speed of 30 km/h. After stopping for 15 minutes at Luton it then travels a further 7 km to Dunstable at an average speed of 28 km/h. Draw a distance-time graph for the journey.
(20) A bus leaves Newton Abbot at 21.20 and travels 25 km to Brixham at an average speed of 20 km/h. After waiting 30 minutes at the Brixham bus station, the bus returns to Newton Abbot at the same speed. Draw a distance-time graph for the return journey.
(21) A heavy lorry leaves Leeds at 9 a.m. and travels south on the M1 at a constant speed of 80 km/h. At 9.15 a.m. a family car leaves Leeds also on the M1, and travelling at a constant speed of 100 km/h. If the distance from Leeds to London is 300 km, draw a distance-time graph showing both journeys and from your graph find when and where the lorry is overtaken.
(22) A lorry leaves Bristol at mid-day to drive to London on the M4 at a constant speed of 64 km/h. At the same time another lorry leaves London for Bristol on the M4, travelling at a constant speed of 60 km/h. If the distance from Bristol to London is 186 km, draw a distance-time graph showing both journeys and from your graph find when and where the lorries pass each other.

Draw a distance-time graph for each of the following:

(23)

Time in s	0	1	2	3	4
Distance in m	0	1.5	3.5	8	20

From your graph, find the instantaneous speed after 3 seconds.

(24)

Time in s	0	5	10	15	20
Distance in m	0	20	50	100	200

From your graph, find the instantaneous speed after 10 seconds.

(25)

Time in mins	0	10	20	30	40
Distance in km	0	4.5	11	21	40

From your graph, find the instantaneous speed after 20 minutes.

Exercise 54b

(1) A car travels 50 km at an average speed of 40 km/h and an average fuel consumption of 10 km/l. It then travels 165 km at an average speed of 60 km/h and an average fuel consumption of 7.5 km/l. Calculate:

(a) the total time taken for the journey

(b) the average speed for the whole journey

(c) the total amount of fuel used

(d) the average fuel consumption for the journey in km/l, correct to 2 significant figures.

(2) At 20.00 a coach leaves Portsmouth and travels 75 km to Brighton at a speed of 60 km/h. After stopping at Brighton for 30 minutes it travels another 35 km to Eastbourne at a speed of 35 km/h. At 21.00 a non-stop coach leaves Eastbourne for Portsmouth at a speed of 55 km/h.

(a) Draw a distance-time graph to show both journeys assuming that both coaches travel at constant speed.

(b) From your graph find where the non-stop coach passes the other.

(c) Calculate the average speed of the stopping coach for the whole journey.

(3) From Preston to Lancaster is 30 km. A man cycles from Lancaster to Preston at a constant speed of 15 km/h, starting at 08.00. Also at 08.00 a car travels from Preston to Lancaster at a constant speed of 60 km/h and after waiting at Lancaster for $\frac{1}{2}$ h, returns to Preston at the same speed.

(a) Draw a distance-time graph showing both journeys.

(b) From your graph find when and where the cyclist and car pass each other for the first time.

(c) Calculate the average speed of the car from the time it leaves Preston to the time it returns.

(4) The table shows the distances travelled by a motorcycle as it starts from rest.

Time in s	0	2	4	6	8	10
Distance in m	0	12.5	30	50	72.5	100

(a) Draw a distance-time graph for this acceleration.

(b) Calculate the average speed of the motorcycle during the first 4 seconds.

(c) Find the instantaneous speed of the motorcycle after 4 s.

(5) The table shows the height of a pellet as it is projected into the air.

Time in s	0	1	2	3	4	5	6
Height in m	0	25	40	45	40	25	0

(a) Draw a distance-time graph for this data.

(b) Find the average speed of the stone in the first 3 seconds.

(c) Find the speed of the stone after 2 seconds.

55 Vectors

Displacement

If a person travels 3 km in one direction and then 4 km in another direction, we can, providing we know the directions, draw a map which shows where he has gone. We can also use the map to decide how far he is from his starting point and which direction he should travel in to get back home by the shortest route.

Example: A boat travels 3 km in a direction 045° and then 4 km in a direction 090°. Find by drawing, how far it is now from its starting point, and the direction to steer in order to return by the shortest route.

The map for the journey is shown below. P is the starting point and Q is the end point.

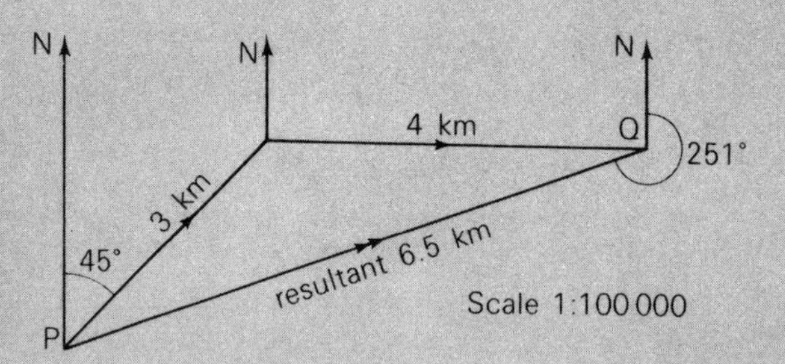

By measurement, PQ = 6.5 km
and $\angle Q = 19°$.
Therefore, course to steer to return to P is 251°.

It is helpful to draw such maps on graph paper. By putting the north–south lines and the east–west lines on the lines of the graph paper, the map becomes easier to draw and is also more accurate.

The amounts moved by the boat (3 km, 4 km) are distances. If the direction is also given (3 km in a direction 045°) then we are told the *displacement*. The line PQ represents the size and direction of the resulting displacement, or *resultant*.

Vector Addition

Some quantities have no direction, but for some quantities, such as displacement, the direction is very important. Here is a list of quantities in which direction is important. An example of each is given in brackets.

displacement (3 km north)
velocity (100 km/h south)
force (5 g upwards)
weight (20 kg downwards).

When the direction of a moving object is given with its speed, the speed and direction together go to make a *velocity*. A force equal to the weight of a 5 g mass is a 5 g force. A weight is a special kind of force. A weight of 20 kg is the force of gravity acting on a 20 kg mass. The size of the force is 20 kg. A quantity such as displacement, velocity or force which has both size and direction is called a *vector* quantity.

All vectors can be added in the same way that displacements are added.

Example: A small aircraft flies at 320 km/h in still air. The pilot sets his course at 080° but is blown off course by an 80 km/h wind blowing from 140°. Find by drawing, the ground speed of the plane and its track.

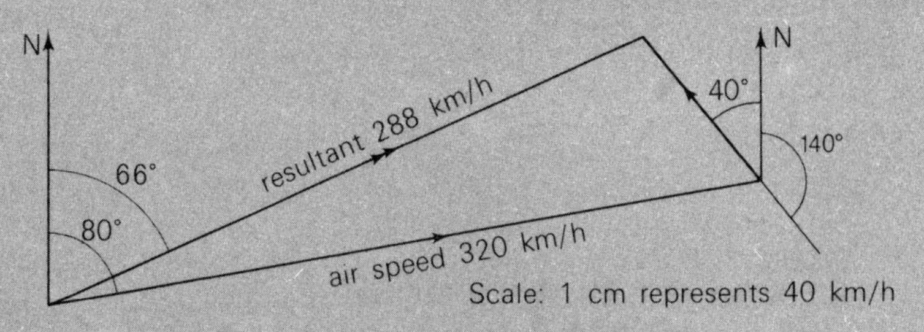

From the drawing, the resultant velocity, or groundspeed is 288 km/h. The track is 66°.

It is important to see that the triangle of vectors in the above example shows velocities only. The distances flown can only be calculated if the time of flight is known.

The direction in which the aircraft points is its *course*. It will only fly in this direction if there is no wind. The speed with which the plane would fly in still air is its *air speed*. The wind direction which is given shows the direction that the wind is blowing from. The speed of the wind is called the *wind speed*. The wind blows the plane off course and the amount that it is blown off course is the *drift*. In the above example the drift is $80° - 66° = 14°$.

The resultant speed of the plane is the speed that it passes over the ground and is called the *ground speed*. The direction that it moves over the ground is called the *track*.

Notice that in order to find the resultant of two vectors, the tail of one vector is 'tacked on' to the head of the other, no matter whether the vectors are displacements, velocities or forces.

Exercise 55a

Solve the following problems by scale drawing. Give all angles to the nearest degree and all other answers to 2 significant figures.

(1) A plane flies 600 km in a direction 030° and 400 km in a direction 070°. Find the resultant displacement.
(2) A ship sails 4 km in a direction 320° and then 3 km in a direction 260°. Find the resultant displacement.
(3) A hiker walks 3 km in a direction 110° and 2 km in a direction 240°. Find the resultant displacement.
(4) A helicopter flies 60 km in a direction 120° and then 100 km due north. Find how far it is from its base and the direction it must fly to get back by the shortest route.
(5) A rambler walks 4 km SW and then 5 km east. Find how far he is from his starting point and the direction that he should take to get back by the shortest route.

In questions 6–10, the course and air speed of a plane are given with the wind speed and direction. Find the ground speed, track and drift in each case.

(6) Course and air speed 320°, 200 km/h. Wind 30 km/h, 270°.
(7) Course and air speed 160°, 160 km/h. Wind 40 km/h, 040°.
(8) Course and air speed 310°, 150 km/h. Wind 50 km/h, 095°.
(9) Course and air speed 255°, 200 km/h. Wind 80 km/h, 105°.
(10) Course and air speed 340°, 300 km/h. Wind 60 km/h, 030°.

In each of questions 11–15, the size and direction of two forces is given. Find the resultant force in each case.

(11) 5 kg at 040° and 5 kg at 160°.
(12) 8 kg at 200° and 10 kg at 130°.
(13) 8 kg at 175° and 7.5 kg at 330°.
(14) 50 kg at 107° and 120 kg at 297°.
(15) 90 kg at 340° and 90 kg at 070°.

Exercise 55b

(1) A yacht sails 6 km in a direction N 25° E from its moorings and then sails 4 km due west.

(a) Make a scale drawing to show the route of the yacht.

(b) From your drawing find the final distance of the yacht from its moorings in km, correct to 2 significant figures.

(c) From your drawing find the final bearing of the yacht from its moorings to the nearest degree.

(d) Calculate, correct to the nearest minute, the time taken for the yacht to return to its moorings if it sails by the shortest route at a constant speed of 20 km/h.

(2) A pilot sets his course at 009° with an airspeed of 180 km/h. If a wind blows in a direction 243° at 50 km/h, draw a vector triangle showing the resultant velocity and from your drawing find

(a) the ground speed of the aircraft in km/h, correct to 2 significant figures

(b) the track of the aircraft to the nearest degree

(c) the drift, to the nearest degree.

(3) A pilot sets his course at 052° with an air speed of 200 km/h. A wind blows him off course so that his ground speed is 170 km/h and the track is 041°.

(a) Draw a vector triangle to scale, showing the course, track, air speed and ground speed.

(b) Calculate the drift and from your drawing, find

(c) the wind speed in km/h, correct to 2 significant figures

(d) the direction of the wind to the nearest degree.

(4) A river flows from west to east. A man sets out in a rowing boat to cross the river by the shortest route, but is carried down stream by a 6 km/h current. If he can row at 3 km/h in still water,

(a) draw a vector triangle to scale and show clearly the scale used and find

(b) the resultant velocity in km/h, correct to 2 significant figures

(c) the angle of drift

(d) the angle that his path makes with the bank.

Answers

Exercise 1a

(1) T	(21) Red-headed Irishmen	(31) 12	(41) Y and M
(2) T	(22) Women athletes	(32) 3	(42) W
(3) F	(23) Bow-legged jockeys	(33) 32	(43) M
(4) T	(24) Absent-minded professors	(34) 11	(44) S
(5) T	(25) White rabbits	(35) 100	(45) W
(6) F	(26) People	(36) 2	(46) Finite
(7) T	(27) School-teachers	(37) 30	(47) Infinite
(8) F	(28) Cutlery	(38) 52	(48) Infinite
(9) F	(29) Our solar system	(39) 366	(49) Finite
(10) F.	(30) Limbs	(40) 100	(50) Infinite

Exercise 1b

(1) (c) People who are neither blind nor deaf
(d) People who are blind but not deaf
(2) (a) $\{1, 2, 3, 4, 5\}$ (d) $\{2, 4, 6, 8, 10\}$
(3) (c) Blue-eyed boys (d) Fair-haired girls without blue eyes
(4) (b) Pupils who take all three subjects (d) 3
(5) (a) 3 (c) 16 (d) 6

Exercise 2a

(1) 21	(10) 5	(19) T	(28) F	(37) 11 736
(2) 41	(11) 20	(20) F	(29) T	(38) 96
(3) 75	(12) 10	(21) T	(30) F	(39) 44
(4) 10	(13) 4	(22) T	(31) 14×40	(40) 573
(5) 12	(14) 17	(23) T	(32) 8×20	(41) 180 000
(6) 6	(15) 21	(24) T	(33) 57×8	(42) 242
(7) 6	(16) T	(25) F	(34) 42×16	(43) 3 000
(8) 30	(17) F	(26) F	(35) 27 183	(44) 2 000
(9) 9	(18) T	(27) T	(36) 4 176	(45) 3

Exercise 2b

(1) (a) 1 144 (b) 72 (c) 160
(2) (a) 4 374 (b) 412 (c) 19 683 (d) 14
(3) (a) 385 km (b) 35 litres (c) 5 l/h (d) 10 l
(4) (a) 320 (b) 2 560 (c) £368
(5) (a) 594 (b) 216 (c) 810 (d) 810

Exercise 3a

(1) $7x$
(2) $24y$
(3) $12z$
(4) $70n$ pence
(5) $2x$ kg
(6) $n + 2$
(7) $30x$ km
(8) $10n$ km
(9) np pence
(10) $x + 1$
(11) $3x + 3y$
(12) $8x + 2y$
(13) $5p - 5q$
(14) $6p - 9q$
(15) $7x^2 + 7$
(16) $10 - 15x^2$
(17) $2x^2y + 2x^2z$
(18) $x^3 + x^3y$
(19) $px - py^2$
(20) $qx^2 + 3qy$
(21) T
(22) T
(23) F
(24) T
(25) F
(26) 17
(27) 13
(28) 42
(29) 12
(30) 36
(31) 16
(32) 6
(33) 3
(34) 13
(35) 3
(36) 36
(37) 25
(38) 7
(39) 72
(40) 50
(41) 10
(42) 10
(43) 27
(44) 49
(45) 1
(46) 12
(47) 0
(48) 0
(49) 25
(50) 5

Exercise 3b

(1) (a) 220 cm (b) 110 m (c) 455
(2) (a) 48 cm (b) 16 cm² (c) 96 cm² (d) 64 cm³
(3) (a) 154 cm² (b) 616 cm² (c) 1 437⅓ cm³
(4) (a) 10 m/s (b) 40 m (c) 40 m
(5) (a) 30 (b) 75

Exercise 4a

(1) 1 950.23
(2) 402.311
(3) 851.264
(4) 1 066.813
(5) 382.758
(6) 12.84
(7) 48.54
(8) 15.755
(9) 19.189
(10) 14.16
(11) 72.667
(12) 137.63
(13) 75.86
(14) 65.87
(15) 152.9
(16) 3 763
(17) 5 270
(18) 16 854
(19) 237 390
(20) 3 040
(21) 2.963
(22) 28.04
(23) 0.5313
(24) 0.0794
(25) 0.0241
(26) 30 469.12
(27) 460.081 68
(28) 520.0205
(29) 2.222 10
(30) 12.685 92
(31) 42.3
(32) 73.81
(33) 103.1
(34) 31.1
(35) 54.9

Exercise 4b

(1) (a) 17.68 (b) 10.88 (c) 48.552 (d) 4.2
(2) (a) 52.90 (b) 132.25 (c) 2.5 (d) 0.4
(3) (a) 117.7 km (b) 22.7 km (c) 286.1 km
(4) (a) 22.4 km (b) 134.4 km (c) 5 577.6 km
(5) (a) 2.34 litres (b) 60.84 litres (c) 20.96 litres (d) 14

Exercise 5a

(1) (a) Acute (b) Reflex (c) Obtuse (d) Obtuse (e) Right angle (f) Reflex

(2) 90°
(3) 30°
(4) 60°
(5) 30°
(6) 120°
(7) 120°
(8) 150°
(9) 90°
(10) 105°
(11) 135°
(12) 105°

(13) (a) T (b) F (c) T (d) T (e) F (f) F
(14) (a) 60° (b) 45° (c) 180° (d) 330° (e) 270° (f) 200°
(15) (a) 020° (N 20° E) (b) 170° (S 10° E) (c) 130° (S 50° E) (d) 220° (S 40° W)
(e) 315° (N 45° W) (f) 230° (S 50° W)

(16) (a) 334° (N 26° W) (b) 250° (S 70° W) (c) 008° (N 8° E) (d) 150° (S 30° E)
(e) 220° (S 40° W) (f) 120° (S 60° E)
(17) N 23° E
(18) 325°
(19) 203° (S 23° W)
(20) 145° (S 35° E)

Exercise 5b

(1) (a) 080° (N 80° E) (b) 260° (S 80° W) (c) 100°
(2) (a) 070° (b) 180° (c) 250°
(3) (b) S 12° E (c) N 8° E
(4) (a) S 37° E (b) 143° (c) 20°

Exercise 6a

(1) £0.25
(2) £0.87
(3) £0.23
(4) £0.07
(5) £0.51½
(6) £1.17
(7) £1.93
(8) £2.51½
(9) £7.03
(10) £5.00½
(11) 63p
(12) 81p
(13) 49½p
(14) 3p
(15) 11p
(16) 126p
(17) 843p
(18) 427½p
(19) 1 192½p
(20) 403p
(21) £9.37
(22) £20.69½
(23) £2.37½
(24) £2 549.73
(25) £156.28½
(26) £3.26
(27) £2.78½
(28) £6.71½
(29) £15.93
(30) £6.82½
(31) £1 370.98
(32) £786.41
(33) £820.54½
(34) £221.59½
(35) £351.32½
(36) £17.23
(37) £32.41
(38) £27.33½
(39) £21.30½
(40) £43.02½
(41) $930.55
(42) DM2.87
(43) L815
(44) 4 990.93f
(45) $43.75
(46) 580.75f
(47) $487.92
(48) £5.20
(49) £63.70
(50) £3.40

Exercise 6b

(1) (a) $74.43, $73.21, $165.70, $154.99 (b) $455.33 (c) $131.67
(2) (a) £2.24 (b) £11.29 (c) £47.74
(3) (a) £49.02 (b) £50.49 (c) a (d) b
(4) (a) 594 km (b) DM30.60 (c) £3.60
(5) (a) L8 134 (b) L1 866 (c) £1.20

Exercise 7a

(1) (−3)
(2) (−1)
(3) (−6)
(4) 3
(5) 2
(6) 4
(7) 0
(8) 0
(9) 0
(10) (−2)
(11) (−1)
(12) (−3)
(13) (−11)
(14) (−10)
(15) (−16)
(16) (−1)
(17) (−3)
(18) (−7)
(19) 6
(20) 2
(21) 3
(22) (−13)
(23) 2
(24) 0
(25) (−1)
(26) (−12)
(27) (−3)
(28) 1
(29) 5
(30) (−15)
(31) 12
(32) 12
(33) 15
(34) (−10)
(35) (−24)
(36) (−35)
(37) (−5)
(38) (−5)
(39) (−5)
(40) (−4)
(41) 4
(42) 5
(43) 4
(44) (−36)
(45) (−24)
(46) 8
(47) (−18)
(48) 8
(49) 2
(50) 8

Exercise 7b

(1) (a) -1 (b) -2 (c) -10 (d) 30 (e) -5
(2) (a) -13 (b) 36 (c) 10 (d) 8
(3) (a) 5 (b) 29 (c) 17 (d) 16
(4) (a) 2 (b) -1 (c) 2 (d) -3
(5) (a) 2 (b) 0 (c) -1 (d) 5

Exercise 8a

(1) (a) Translation (b) Rotation (c) Translation, Reflection (d) Rotation, Reflection (e) Rotation (f) Rotation
(3) (a) 6 (b) 3 (c) 5 (d) 2 (e) 6 (f) 4
(4) a, b, c, e, f, g
(5) a, d, e, f

Exercise 8b

(1) (a) Translation (c) 2 (d) Yes
(2) (a) Translation (b) none (c) Yes (d) Yes
(3) (a) Rotation (c) No (d) 8
(4) (a) Translation, rotation, reflection (b) 6 (c) 6 (d) Yes

Exercise 9a

(1) $\{1, 2, 3, 4, 6, 9, 12, 18, 36\}$
(2) $\{1, 2, 4, 8, 16, 32\}$
(3) $\{1, 2, 4, 5, 8, 10, 20, 40\}$
(4) $\{1, 2, 4, 7, 8, 14, 28, 56\}$
(5) $\{1, 2, 5, 10, 25, 50\}$
(6) $\{1, x, y, z, xy, xz, yz, xyz\}$
(7) $\{1, a, b, a^2, ab, a^2b\}$
(8) $\{1, a, b, ab, b^2, ab^2\}$
(9) $\{1, x, a + b, x(a + b)\}$
(10) $\{1, x, y, x^2, y^2, xy^2, x^2y, x^2y^2\}$
(14) $\{1, 2, 3, 6\}$
(16) $\{1, 2, 3, 4, 6, 8, 12, 24\}$
(17) $\{2, 4, 6, 8, 10\}$
(18) $\{5, 10, 15, 20, 25, 30\}$
(19) $\{10, 20, 30, 40, 50\}$
(20) $\{7, 14, 21, 28, 35, 42, 49\}$
(21) T
(22) F
(23) F
(24) T
(25) T
(26) T
(27) F
(28) F
(29) T
(30) F
(31) $2 \cdot 3 \cdot 5 \cdot 7$
(32) $3 \cdot 5 \cdot 13$
(33) $2 \cdot 3 \cdot 7^2$
(34) $2 \cdot 3 \cdot 5 \cdot 11^2$
(35) $2 \cdot 3^2 \cdot 5^2 \cdot 7$
(36) $2^3 \cdot 3 \cdot 5$
(37) $3^4 \cdot 5 \cdot 7$
(38) $2^3 \cdot 3^3$
(39) $2 \cdot 3^6$
(40) $2 \cdot 3 \cdot 5^4$
(41) 30
(42) 35
(43) 28
(44) 36
(45) 84
(46) 45
(47) 66
(48) 52
(49) 51
(50) 46

Exercise 9b

(1) (a) $\{1, 2, 3, 4, 5, 6, 10, 12, 15, 20, 30, 60\}$ (b) $2^2 . 3 . 5$ (d) 15
(2) (a) $\{1, 2, 3, 4, 6, 8, 12, 16, 24, 48\}$ (b) $2^4 . 3$ (c) 96, 144, 48 (d) 144
(3) (a) $\{1, a, b, c, ab, ac, bc, a^2, a^2b, a^2c, a^2bc\}$ (b) $\{1, a, b, c, ab, ac, bc, c^2, ac^2, bc^2, abc^2\}$ (c) a^2bc^2
(4) (a) $2^2 \cdot 3^2 \cdot 7^2$ (b) $2^4 \cdot 5 \cdot 7^2$ (c) 196 (d) 42
(5) (a) $3^4 \cdot 7^4$ (b) 441 (c) 21

Exercise 10a

(1) x^7
(2) x^7
(3) y^{13}
(4) p^7
(5) a^{11}
(6) x^9
(7) a^{11}
(8) $10x^7$
(9) $18x^7$
(10) $8a^5$
(11) $6x^3y^3$
(12) $8x^4y^5$
(13) $6a^7b^6$
(14) $18x^{12}y^6$
(15) $8x^8y^5z^2$
(16) x^2
(17) a^2
(18) 1
(19) $2x^4$
(20) $5y^6$
(21) $2xy^2$
(22) $5x^3$
(23) $4xy^3$
(24) $7a^2b^8c$
(25) $7p^8q^5r^2$
(26) x^6
(27) a^4b^6
(28) $4p^4q^2$
(29) $9x^4y^6$
(30) $25a^6b^2c^4$
(31) $9x^4y^{10}z^2$
(32) $81a^{10}b^{10}c^4$
(33) $3ab^2$
(34) $2x^2y^3$
(35) $4pq^4r$
(36) $6l^2m^5$
(37) $8xy^4$
(38) $7a^3b^3$
(39) $5a^6b^3$
(40) $9p^2q^3$

Exercise 10b

(1) (a) $54x^3y^4z^6$ (b) $6xz^2$ (c) $324x^4y^4z^8$ (d) $9x^2y^4z^4$
(2) (a) $108a^7b^{11}c^3$ (b) $12a^5b^5c$ (c) $9a^2b^6c^2$ (d) $6a^3b^4c$
(3) (a) $180p^3q^3r^4$ (b) $5p^3qr^2$ (c) $4p^4q^4r^2$ (d) $2p^4q^3r$
(4) (a) $36x^4y^6z^8$ (b) $1\,296x^8y^{12}z^{16}$ (c) $6x^2y^3z^4$ (d) $9x^2y^2z^2$
(5) (a) $64a^2b^4c^2$ (b) $4a^2c^2$ (c) $256a^4b^4c^4$ (d) $256a^4b^4c^4$ (e) $2b$

Exercise 11a

(1) £50
(2) £72
(3) £38
(4) £198
(5) £114.57
(6) £107
(7) £85
(8) £93
(9) £67
(10) £213
(11) £480 690
(12) £508 750
(13) £804 930
(14) £750 876
(15) £679 715
(16) £1 240
(17) £6 825
(18) £1 140
(19) £2 048
(20) £1 474
(21) £750
(22) £1 675
(23) £1 030
(24) £1 305
(25) £2 190
(26) £953.10
(27) £651.75
(28) £1 614.08
(29) £1 479.32
(30) £586.50

Exercise 11b

(1) (a) £210.60 (b) £10.53 (c) £341 250 (d) £5 250
(2) (a) £7 600 (b) £554 800 (c) £106 400
(3) (a) £1 260 (b) £3 150 (c) £1 102.50 (d) 25%
(4) (a) £2 050 (b) £676.50 (c) £435.48 (d) £1 111.98
(5) (a) £46.56 (b) £5.82 (c) £165.55 (d) £217.93

Exercise 12a

(1) $3x + 15$
(2) $7x - 14$
(3) $3x + 12$
(4) $8x - 12$
(5) $6x + 21$
(6) $10x - 6$
(7) $2x^2 + 2x$
(8) $3x^2 - 6x$
(9) $4x^2 + 7x$
(10) $6x^2 + 12x$
(11) $-2x^2 - 12x$
(12) $-9x^2 + 21x$
(13) $4x^2 + 4xy$
(14) $-3x - 2xy$
(15) $b - a$
(16) $7x + 17$
(17) $7x + 17$
(18) $16x + 3$
(19) $13x + 12$
(20) $8x^2 - 6x + 3$
(21) $6x^2 + 8$
(22) $4x^2 + 13x - 3$
(23) $10x^2 - 8x$
(24) $3a + 27$
(25) $-2p + 24$
(26) $6x^2 + 3x + 3$
(27) $2x^2 - 6x - 12$
(28) $-x^2 - 5x$
(29) $8x^2 + 17x$
(30) $4x^2 - 13x - 6$
(31) $x^2 + 4x + 3$
(32) $2x^2 + 10x + 8$
(33) $3x^2 + 18x + 15$
(34) $6x^2 + 15x + 6$
(35) $4x^2 - 5x - 6$
(36) $4x^2 - 2x - 12$
(37) $2x^2 + x - 6$
(38) $6x^2 + 7x - 5$
(39) $x^2 - 1$

(40) $3x^2 + x - 2$
(41) $8x^2 - 12x - 20$
(42) $8x^2 - 10x - 3$
(43) $3x^2 - 21x + 18$
(44) $5x^2 - 17x + 6$
(45) $3x^2 - 25x + 28$

Exercise 12b

(1) (a) $9x + 7$ (b) $2x + 3$ (c) $x^2 + 4x + 3$ (d) $6x^2 - 5x - 4$
(2) (a) $-15x - 9$ (b) $7x - 4$ (c) $8x + 13$ (d) $10x^2 - 29x - 21$
(3) (a) $-4x + 6$ (b) $x + 4$ (c) $6x^2 - 7x - 3$ (d) $9x^2 + 6x + 1$
(4) (a) $5x + 1$ (b) $-x + 10$ (c) $6x^2 - x - 12$ (d) $9x^2 + 24x + 16$
(5) (a) $4x$ (b) $4x - 4$ (c) $4x^2 - 1$ (d) $4x^2 - 4x + 1$

Exercise 13a

(1) (a) 54° (b) 60° (c) 20° (d) 80° (e) 30° (f) 120° (g) 30° (h) 130°
(2) 110°
(3) 10°
(4) 60°
(5) 70°
(6) 90°
(7) 90°
(8) 120°
(9) 90°
(10) 90°

Exercise 13b

(1) (a) 110° (b) Obtuse-angled (c) Scalene (d) AB
(2) (a) 123° (b) 123° (c) 57° (d) 123°
(3) (a) 61° (b) 119° (c) 61°
(4) (a) 55° (b) 93° (c) 55°
(5) (a) 130° (b) 60° (c) 65°

Exercise 14a

(1) Wheat
(2) Oats
(3) 20
(4) 2 million
(5) 17 million
(6) 1960–70
(7) N. Ireland
(8) 33%
(9) 11–17 years
(10) 15–16 years
(11) 13–14 years
(12) Thursday
(13) Saturday
(14) Saturday
(15) Not known

Exercise 14b

(1) (a) Pictogram (b) 100 million (c) 1 500 million (d) 60
(2) (b) 508 mm (c) May–July (d) December–February
(3) (b) 10° (c) 160°
(4) (b) 10–11 years (c) 146 cm
(5) (b) 10° C (c) Not known

Exercise 15a

(1) 11, 12, 13
(2) 22, 26, 30
(3) −4, −6, −8
(4) 0, −5, −10
(5) 32, 64, 128
(6) 48, 96, 192
(7) 162, 486, 1 458
(8) $\frac{1}{4}, \frac{1}{8}, \frac{1}{16}$
(9) 13, 21, 34
(10) 15, 21, 28
(11) 25, 36, 49
(12) 125, 216, 343
(13) 30, 42, 56
(14) $\frac{5}{6}, \frac{6}{7}, \frac{7}{8}$
(15) $\frac{9}{11}, \frac{11}{13}, \frac{13}{15}$
(16) 31, 43, 57
(17) 81, 243, 729
(18) $\frac{1}{32}, \frac{1}{64}, \frac{1}{128}$
(19) 5, 1, $\frac{1}{5}$
(20) 1, 2, 4
(25) 55
(26) 11, 13, 17, 19, 23, 29
(27) 2, 4, 8, 16, 32, 64, 128, 256
(28) 3, 9, 27, 81, 243, 729
(29) 4, 16, 64, 256
(30) 5, 25, 125, 625

Exercise 15b

(1) (a) 224 (b) 8 (c) 32 (d) 256
(2) (b) 1, 3, 6, 10, 15, 21, 28, 36, 45, 55
(3) (a) 1, 5, 10, 10, 5, 1 (b) 11th (c) 512
(4) (a) $5^2 - 4^2 = 9$, $6^2 - 5^2 = 11$, $7^2 - 6^2 = 13$ (b) $20^2 - 19^2 = 39$ (c) 245

Exercise 16a

(1) 52°
(2) 108°
(3) 6.2 cm
(4) 8.8 cm
(5) 3.3 cm
(6) 5.8 cm
(7) 11.8 cm
(8) 68°
(9) 30°
(10) 8.0 cm
(11) 6.9 cm
(12) 4 cm
(13) 116°
(14) 65°
(15) 10.0 cm
(16) 3.6 cm
(17) 45°
(18) 90°
(19) 4.5 cm
(20) 6.0 cm
(21) 10.0 cm
(22) 5.2 cm
(23) 11.2 cm
(24) 8.5 cm
(25) 110°

Exercise 16b

(1) (b) 40° (d) 3.2 cm
(2) (b) 3.6 cm (d) 6.5 cm
(3) (b) 78° (d) 2.3 cm
(4) (b) 6.0 cm (d) 90°
(5) (b) 10.0 cm (d) 2.8 cm

Exercise 17a

(1) 302 cm
(2) 1 405 cm
(3) 16.5 cm
(4) 1.73 m
(5) 3 154 m
(6) 7.36 m
(7) 0.1362 m^2
(8) 0.2536 m^2
(9) 32 000 m^2
(10) 1.7 ha
(11) 3 500 ha
(12) 430 ha
(13) 3.0 l
(14) 5.2 l
(15) 0.7 l
(16) 4 000 ml
(17) 2 500 ml
(18) 300 ml
(19) 5.0 g
(20) 3 200 g
(21) 4 700 g
(22) 7.5 kg
(23) 4.62 kg
(24) 5 300 kg
(25) 11.46 m
(26) 40.15 cm
(27) 240.2 mm
(28) 13.859 km
(29) 3.78 m^2
(30) 6.03 ha
(31) 21.87 km^2
(32) 28.431 l
(33) 15.278 kg
(34) 56.751 g
(35) 25.008 t

Exercise 17b

(1) (a) 2.7 m^2 (b) 3.2 t (c) 54 (53.5)
(2) (a) 350 m (b) 0.75 ha (c) £1 650
(3) (a) 150 km (b) 12.5 l (c) 396 km (d) £5.28
(4) (a) 5 544 cm^2 (b) 38 808 cm^3 (c) 345 kg

Exercise 18a

(1) $\frac{4}{7}, \frac{3}{5}, \frac{2}{3}$
(2) $\frac{2}{5}, \frac{3}{7}, \frac{5}{11}$
(3) $\frac{5}{8}, \frac{2}{3}, \frac{3}{4}$
(4) $\frac{1}{6}, \frac{1}{5}, \frac{2}{9}$
(5) $\frac{3}{4}, \frac{4}{5}, \frac{7}{8}, \frac{9}{10}$
(6) $\frac{4}{11}, \frac{3}{7}, \frac{2}{3}, \frac{7}{8}$
(7) $\frac{4}{7}, \frac{5}{8}, \frac{2}{3}, \frac{3}{4}$
(8) $\frac{1}{3}, \frac{2}{5}, \frac{5}{9}, \frac{4}{7}$
(9) $\frac{3}{7}, \frac{1}{2}, \frac{3}{4}, \frac{4}{5}$
(10) $\frac{2}{3}, \frac{5}{8}, \frac{7}{10}, \frac{6}{7}$
(11) $6\frac{1}{6}$
(12) $5\frac{15}{28}$
(13) $5\frac{13}{15}$
(14) $9\frac{8}{35}$
(15) $10\frac{7}{12}$
(16) $2\frac{1}{10}$
(17) $1\frac{13}{28}$
(18) $1\frac{7}{9}$
(19) $2\frac{14}{15}$
(20) $\frac{13}{18}$
(21) $2\frac{11}{12}$
(22) $4\frac{2}{5}$
(23) $3\frac{7}{30}$
(24) $5\frac{5}{12}$
(25) $3\frac{9}{28}$
(26) $7\frac{53}{60}$
(27) $2\frac{11}{12}$
(28) $2\frac{23}{24}$
(29) $5\frac{23}{24}$
(30) $5\frac{5}{24}$
(31) $\frac{10}{21}$
(32) $\frac{3}{10}$
(33) $\frac{8}{45}$
(34) $\frac{2}{5}$
(35) $\frac{1}{6}$
(36) 4
(37) $3\frac{1}{8}$
(38) $5\frac{1}{2}$
(39) 6
(40) $1\frac{1}{2}$

(41) $1\frac{1}{3}$
(42) $2\frac{5}{8}$
(43) $\frac{9}{10}$
(44) $\frac{3}{11}$
(45) $\frac{3}{4}$
(46) $2\frac{4}{9}$
(47) 7
(48) $\frac{9}{10}$
(49) $2\frac{1}{7}$
(50) $3\frac{9}{11}$

Exercise 18b

(1) (a) $2\frac{13}{20}$ (b) $3\frac{1}{8}$ (c) $1\frac{5}{9}$
(2) (a) $\frac{7}{20}, \frac{3}{7}, \frac{5}{8}$ (b) $2\frac{5}{12}$ (c) $\frac{1}{8}$
(3) (a) $4\frac{1}{2}$ (b) $1\frac{4}{7}$ (c) $1\frac{3}{17}$
(4) (a) $4\frac{1}{6}$ (b) $1\frac{1}{3}$ (c) $3\frac{1}{8}$

Exercise 19a

(1) $\frac{3a + 2b}{6}$
(2) $\frac{5x + 4y}{20}$
(3) $\frac{2b + 3a}{ab}$
(4) $\frac{3y - 2x}{xy}$
(5) $\frac{b - 2}{ab}$
(6) $\frac{3 + 4b}{b^2}$
(7) $\frac{-5}{4a}$
(8) $\frac{3 + a}{3}$
(9) $\frac{b - 2}{b}$
(10) $\frac{3x^2 - 4}{x^2}$
(11) $\frac{2ab + b + 3a}{ab}$
(12) $\frac{5y - 2}{y}$
(13) $\frac{4 + 3x}{4}$
(14) $\frac{6 + 3x}{2x}$
(15) $\frac{2c + a - 3b}{abc}$
(16) $\frac{1}{7}$
(17) $\frac{5}{8}$
(18) $1\frac{1}{11}$
(19) $13\frac{3}{4}$
(20) $6\frac{7}{8}$
(21) $3\frac{1}{3}$
(22) $4\frac{25}{28}$
(23) $3\frac{3}{4}$
(24) $4\frac{23}{24}$
(25) $1\frac{1}{2}$
(26) $1\frac{1}{3}$
(27) $1\frac{1}{2}$
(28) $\frac{9}{14}$
(29) $\frac{20}{21}$
(30) $1\frac{1}{4}$

Exercise 19b

(1) (a) $\frac{4x - 3y}{12}$ (b) $\frac{12 - xy}{4x}$ (c) $\frac{x^2 + 2x + 3}{x^2}$
(2) (a) $2\frac{3}{7}$ (b) $1\frac{11}{16}$ (c) $6\frac{1}{8}$
(3) (a) $6\frac{9}{16}$ (b) $5\frac{3}{8}$ (c) $\frac{1}{7}$
(4) (a) $1\frac{1}{12}$ (b) $6\frac{1}{2}$ (c) $\frac{1}{6}$
(5) (a) $-\frac{5}{7}$ (b) $2\frac{1}{3}$ (c) $-\frac{15}{49}$

Exercise 20a

(1) (a) 3.5 cm (b) 10 cm^2
(2) 50 cm
(3) (a) 15 m (b) 44 cm^2
(4) 1 m^2
(5) $\frac{1}{400}$
(6) $\frac{1}{100\,000}$
(7) 5.5 km
(8) 1.5 m^2
(9) 40 000
(10) (a) 1 km^2 (b) 1 ha

Exercise 20b

(1) (a) 9.6 m^2 (b) 75.9 m^2
(2) (a) 62 cm (b) Curved, 4 m
(3) (a) 1.5 m (b) 94.2 m^2
(4) (a) 72 m^2 (b) 7.5 m^2 (c) 20 cm
(5) (a) 28 m^2 (b) $\frac{1}{2}$

Exercise 21a

(1) 46_{10}
(2) 92_{10}
(3) 53_{10}
(4) 151_{10}
(5) 326_{10}
(6) 73_{10}
(7) 70_{10}
(8) 143_{10}
(9) 170_{10}
(10) 198_{10}
(11) 5_{10}
(12) 6_{10}
(13) 11_{10}
(14) 27_{10}
(15) 52_{10}
(16) 150_{10}
(17) 279_{10}
(18) 298_{10}
(19) 311_{10}
(20) 419_{10}
(21) 44_{10}
(22) 77_{10}
(23) 235_{10}
(24) 340_{10}
(25) 433_{10}
(26) 333_5
(27) 441_5
(28) $1\,120_5$
(29) $3\,010_5$
(30) $10\,011_5$
(31) $1\,111_2$
(32) $10\,011_2$

(33) $100\,101_2$
(34) $111\,100_2$
(35) $1\,001\,011_2$
(36) $2\,201_3$
(37) $10\,102_3$
(38) $12\,100_3$
(39) $20\,000_3$
(40) $12\,212_3$
(41) 45_7
(42) 60_7
(43) 111_7
(44) 125_7
(45) 202_7
(46) 53_{12}
(47) 73_{12}
(48) 106_{12}
(49) 202_{12}
(50) 212_{12}
(51) $1\,205_6$
(52) 717_8
(53) 241_5
(54) $10\,110_2$
(55) 433_7
(56) 331_5
(57) $42\,100_5$
(58) $4\,160_7$
(59) $1\,101_2$
(60) $46T_{12}$

Exercise 21b

(1) (a)

+	1	2	3	4
1	2	3	4	10
2	3	4	10	11
3	4	10	11	12
4	10	11	12	13

(b) $4\,111_5$ (c) 402_5 (d) $34\,210_5$

(2) (a)

+	1	2	3	4	5	6	7
1	2	3	4	5	6	7	10
2	3	4	5	6	7	10	11
3	4	5	6	7	10	11	12
4	5	6	7	10	11	12	13
5	6	7	10	11	12	13	14
6	7	10	11	12	13	14	15
7	10	11	12	13	14	15	20

(b) $1\,144_8$ (c) 567_8 (d) $1\,162_8$

(3) (a)

×	1	2	3
1	1	2	3
2	2	10	12
3	3	12	21

(b) $11\,303_4$ (c) $33\,000_4$ (d) $1\,203_4$

(4) (a)

×	1	2	3	4	5
1	1	2	3	4	5
2	2	4	10	12	14
3	3	10	13	20	23
4	4	12	20	24	32
5	5	14	23	32	41

(b) $1\,143_6$ (c) 352_6 (d) $2\,540_6$

(5) (a) $1\,174_8$ (b) 711_{10} (c) $100\,100_2$ (d) 22_3 (e) $1\,611_7$

Exercise 22a

(1) 43%
(2) 22%
(3) 4%
(4) 123%
(5) 203%
(6) 40%
(7) 30%
(8) 35%
(9) 6%
(10) 28%
(11) 7.5%
(12) 37.5%
(13) 31.25%
(14) 280%
(15) 235%
(16) 75%
(17) 35%
(18) 1%
(19) 267%
(20) 543%
(21) 0.35
(22) 0.21
(23) 0.43
(24) 1.21
(25) 2.35
(26) 0.4
(27) 0.75
(28) 0.625
(29) 0.375
(30) 0.6
(31) 0.43
(32) 0.83
(33) 0.64
(34) 0.44
(35) 0.67
(36) 0.714
(37) 0.667
(38) 0.167
(39) 0.273
(40) 0.778
(41) 168
(42) 146 days
(43) £13.76
(44) 9.03 km
(45) 126°
(46) 365 m^2
(47) 900 ml
(48) 18 mins
(49) 3 m^3
(50) 126°
(51) 9%
(52) 24%
(53) 35%
(54) 28%
(55) 32%
(56) 15%
(57) 20%
(58) 23%
(59) 28%
(60) 60%

Exercise 22b

(1) (a) 75% (b) 45% (c) 33.75% (d) £5.40
(2) (a) 53% (b) 96 (c) 20% (d) 94.5%
(3) (a) £227.80 (b) £11.39 (c) 2.5% (d) 20%
(4) (a) 400 000 t (b) 20% (c) 150 000 t (d) 12%
(5) (a) 12 million (b) 14% (c) 78 million (d) 52%

Exercise 23a

(1) 1.7302
(2) 1.6217
(3) 0.8646
(4) 0.6907
(5) 2.5696
(6) 2.6733
(7) 2.9292
(8) 3.9732
(9) 3.8710
(10) 2.3962
(11) $\bar{1}.7933$
(12) $\bar{1}.6172$
(13) $\bar{2}.4998$
(14) $\bar{1}.6140$
(15) $\bar{3}.5070$
(16) 516.4
(17) 64.77
(18) 1 102
(19) 130.7
(20) 2.621
(21) 8.855
(22) 6.794
(23) 47 200
(24) 202.8
(25) 65.52
(26) 0.4181
(27) 0.6522
(28) 0.074 66
(29) 0.002 581
(30) 0.4478
(31) 123
(32) 173
(33) 286
(34) 305
(35) 126
(36) 35.6
(37) 284
(38) 2 850
(39) 473
(40) 1 550
(41) 13.1
(42) 26.1
(43) 28.0
(44) 5.38
(45) 8.06
(46) 0.742
(47) 0.815
(48) 0.033
(49) 0.219
(50) 0.0529

Exercise 23b

(1) (a) 2 110 (b) 1.25 (c) 0.802 (d) 1 390
(2) (a) 160 (b) 0.001 34 (c) 0.0214 (d) 1 880
(3) (a) 724 (b) 4.16 (c) 24.5 (d) 59.2
(4) (a) 14.6 (b) 17.0 (c) 213 (d) 248

Exercise 24a

(1) 5
(2) -9
(3) $\frac{1}{2}$
(4) $1\frac{1}{2}$
(5) $-\frac{2}{3}$
(6) 3
(7) 3
(8) $\frac{1}{2}$
(9) -1
(10) $-\frac{1}{4}$
(11) 2
(12) 5
(13) -1
(14) $-\frac{2}{3}$
(15) -1
(16) $\frac{2}{3}$
(17) $\frac{1}{5}$
(18) $\frac{1}{4}$
(19) $\frac{2}{3}$
(20) $\frac{3}{5}$
(21) 4
(22) 6
(23) 2
(24) $-1\frac{1}{2}$
(25) -1
(26) 7
(27) 5
(28) $-\frac{1}{2}$
(29) 5
(30) $\frac{1}{2}$
(31) 5
(32) 3
(33) 1
(34) 11
(35) 6
(36) -1
(37) 3
(38) -2
(39) $\frac{1}{2}$
(40) 0
(41) 2
(42) 5
(43) 8 years
(44) 7 cm
(45) £9

Exercise 24b

(1) (a) 28 (b) 4 (c) 4 (d) 2 km
(2) (a) -21 (b) $\frac{4}{5}$ (c) 3 (d) 5 years
(3) (a) 2 (b) 7 (c) 2 (d) 7
(4) (a) 0 (b) 2 (c) 2 (d) 18

Exercise 25a

(1) A circle.
(2) The surface of a sphere.
(3) The perpendicular bisector of PQ.
(4) A plane which bisects PQ.
(5) Two concentric circles.
(7) A circle and all the points inside it.
(8) A sphere radius 1 million km.
(9) The arc of a circle.

Exercise 25b

(1) (b) All 5 cm. A square. (c) The perpendicular bisector of AB. (d) The straight line through P and Q.
(2) (d) All points on KL and MN.
(3) (b) 10 cm (d) the circumcircle of $\triangle$ABC.
(4) (b) Circle (c) Smaller (d) $n \to n + 18$
(5) (a) Circle (b) straight line (c) arc of a circle (d) arc of a circle.

Exercise 26a

(1) 720°
(2) 1 080°
(3) 1 800°
(4) 2 700°
(5) 45°
(6) 36°
(7) 18°
(8) 10°
(9) 40°
(10) 9°
(11) 30°
(12) 12°
(13) 135°
(14) 120°
(15) 150°
(16) 108°
(17) 5.9 cm
(18) 3.8 cm
(19) 3.1 cm
(20) 2.6 cm

Exercise 26b

(1) (a) Pentagon (b) 18° (c) 162° (d) 3.8 cm
(2) (a) Decagon (b) 540° (c) 10°
(3) (a) 8 (b) $(2n - 4)$ right angles (c) 40° (d) 140°
(4) (a) Square (b) 120° (c) Hexagon (d) 4.7 cm

Exercise 27a

(1) x^{-4}
(2) x^{-5}
(3) $x^0 = 1$
(4) x^{-4}
(5) y^{-4}
(6) p^{-2}
(7) $z^0 = 1$
(8) x^{-3}
(9) q^{-5}
(10) s^{-5}
(11) x^2
(12) y^{-2}
(13) s^4
(14) x^{-3}
(15) $z^0 = 1$
(16) x^5
(17) x^7
(18) $x^{\frac{3}{2}}$
(19) $x^{\frac{3}{2}}$
(20) $x^0 = 1$
(21) x^{-2}
(22) y^{-4}
(23) p^{-9}
(24) z^{-8}
(25) a^{-3}
(26) $\dfrac{1}{x^6}$
(27) $\dfrac{1}{y^3}$
(28) $\dfrac{1}{z}$
(29) $\dfrac{1}{p^7}$
(30) $\dfrac{1}{z^4}$
(31) $\frac{1}{11}$
(32) $\frac{1}{32}$
(33) $\frac{1}{144}$
(34) 1
(35) 3
(36) 2
(37) $\frac{1}{81}$
(38) 1
(39) 5
(40) 1
(41) $2x^2$
(42) $3x^3$
(43) $8x^2$
(44) $9x^4$
(45) $5x^3$
(46) $3 \times 10^3 = 3\,000$
(47) $2 \times 10^2 = 200$
(48) $5 \times 10^4 = 50\,000$
(49) $4 \times 10 = 40$
(50) $3 \times 10^6 = 3\,000\,000$
(51) 2.7×10^6
(52) 3.1×10^3
(53) 4.2×10^7
(54) 5.7×10^2
(55) 9.4×10^6
(56) 3.4×10^{-1}
(57) 8.7×10^{-3}
(58) 3.2×10^{-6}
(59) 4.6×10^{-4}
(60) 9.1×10^{-8}

Exercise 27b

(1) (a) 288 (b) $5x^3y^5$ (c) $3x^{-3}y$ (d) 3.2×10^2 m^2
(2) (a) 637 (b) 4 (c) 1.0×10^{-2} (d) 1
(3) (a) $20x^2 + 22x + 6$ (b) -1 (c) 784 (d) $\frac{1}{64}$
(4) (a) 3 (b) 5 (c) $15x^{-2}y^7$ (d) $\frac{4}{5}xy^{-2}$
(5) (a) 2.3×10^3 km (b) 1.7×10^3 km (c) 208° (S 28° W) (d) $4\frac{1}{4}$ hours.

Exercise 28a

(1) 2.387
(2) 1.789
(3) 3.082
(4) 2.170
(5) 2.973
(6) 2.592
(7) 2.713
(8) 2.381
(9) 2.987
(10) 3.074
(11) 6.856
(12) 9.434
(13) 5.657
(14) 5.992
(15) 6.943
(16) 8.044
(17) 3.178
(18) 7.281
(19) 6.541
(20) 9.714
(21) 2.506
(22) 7.727
(23) 6.468
(24) 2.147
(25) 2.373
(26) 5.685
(27) 9.026
(28) 1.948
(29) 8.203
(30) 9.830
(31) 23.72
(32) 96.58
(33) 18.51
(34) 21.26
(35) 55.0
(36) 22.84
(37) 90.34
(38) 57.21
(39) 29.02
(40) 31.0
(41) 63.56
(42) 0.1849
(43) 0.2866
(44) 0.2902
(45) 0.386
(46) 0.1034
(47) 0.2035
(48) 0.2486
(49) 0.4889
(50) 0.2858

Exercise 28b

(1) (a) $7ab^2$ (b) 0.2302 (c) 14.1 (d) 21.7 m
(2) (a) 8×10^3 (b) 192.9 (c) 48.37 (d) 45.98 cm
(3) (a) $9x^3y^4$ (b) 6×10^{-3} (c) 13.9 (d) 17.9 cm
(4) (a) 0.5692 (b) 11.4 (c) 12.3 cm² (d) 3.51 cm
(5) (a) $3x^{-2}$ (b) {2, 3, 7} (c) 42 (d) 7.8 cm

Exercise 29a

(1) $\bar{4}.9$
(2) $\bar{5}.6$
(3) 1.7
(4) 2.7
(5) 2.8
(6) $\bar{4}.5$
(7) $\bar{1}.8$
(8) $\bar{1}.6$
(9) $\bar{1}.8$
(10) 2.1
(11) 2.2
(12) 2.4
(13) $\bar{1}.6$
(14) $\bar{3}.1$
(15) $\bar{1}.1$
(16) $\bar{3}.1$
(17) $\bar{2}.3$
(18) $\bar{3}.0$
(19) $\bar{2}.2$
(20) 0.6
(21) $\bar{5}.2$
(22) $\bar{1}.3$
(23) $\bar{6}.2$
(24) $\bar{5}.2$
(25) 5.2
(26) 4.1
(27) 6.3
(28) 5.4
(29) 1.2
(30) 1.3
(31) 1.2
(32) $\bar{2}.1$
(33) $\bar{2}.4$
(34) $\bar{4}.3$
(35) 2.8
(36) 4.7
(37) 5.7
(38) 2.5
(39) $\bar{4}.7$
(40) $\bar{3}.9$
(41) $\bar{4}.2$
(42) $\bar{7}.8$
(43) 0.7
(44) 1.8
(45) $\bar{1}.7$
(46) $\bar{2}.5$
(47) $\bar{3}.7$
(48) $\bar{4}.3$
(49) $\bar{4}.8$
(50) $\bar{2}.7$
(51) 2.60
(52) 0.114
(53) 1.76
(54) 0.085 0
(55) 0.016 5
(56) 168
(57) 3.05
(58) 0.004 65
(59) 23.7
(60) 0.012 5

Exercise 29b

(1) (a) 14.0 (b) 17.0 (c) 1.19 (d) 1.33
(2) (a) 0.0489 (b) 1 100 (c) 2.85 (d) 0.0518 cm³
(3) (a) 3.04 (b) 0.123 (c) 0.119 m² (d) 31.1 km
(4) (a) 0.329 (b) 8.81 (c) 0.680 cm² (d) 0.0755 m²
(5) (a) 62.0 (b) £4.76 (c) 6.96 (d) 4.58 m

Exercise 30a

(6)

solid	F	V	E
tetrahedron	4	4	6
cube	6	8	12
octahedron	8	6	12
dodecahedron	12	20	30
icosahedron	20	12	30

(7) Yes
(8) Cube and octahedron. Cube has 6 faces and 8 vertices. Octahedron has 8 faces and 6 vertices.
(9) Dodecahedron and icosahedron. Dodecahedron has 12 faces and 20 vertices. Icosahedron has 20 faces and 12 vertices.
(10) Tetrahedron. It has the same number of faces and vertices.
(15) No

Exercise 30b

(1) (a) 8 (b) 12 (c) 36 cm (d) 31.2 cm^2
(2) (a) 20 (b) 30 (c) 34 cm^2
(3) (a) Dodecahedron (b) regular pentagon (c) cube (d) octahedron
(4) (a) 2 cm (b) 8 cm^3 (c) 96 cm^2 (d) 64 cm^3
(5) (a) tetrahedron (b) cone (c) hemisphere (d) sphere

Exercise 31a

(5) 2
(6) $\frac{1}{2}$
(7) $\frac{3}{2}$
(8) 3
(9) -2
(10) -2

Exercise 31b

(1) (b) (i) 3 (ii) 2 (c) 1
(2) (b) (i) -4 (ii) $2\frac{1}{2}$ (c) -2
(3) (b) 2 (d) (2, 3)
(4) (b) 3 (c) (1, 0) (d) $1\frac{1}{2}$ square units

Exercise 32a

(1) 8.54 cm
(2) 11.4 cm
(3) 12.5 m
(4) 12.4 km
(5) 13.2 km
(6) 9.75 cm
(7) 10.6 cm
(8) 9.38 m
(9) 19.2 m
(10) 22.4 cm
(11) 11.7 cm
(12) 17.5 cm
(13) 9.64 cm
(14) 9.80 cm
(15) 4.24 m
(16) 21.2 cm
(17) 8.60 cm
(18) 11.3 cm
(19) 5.20 cm
(20) 2.83 cm

Exercise 32b

(1) (a) 5.83 cm (c) 5.8 cm (d) 30 cm^2
(2) (a) 7.62 cm (b) 11.4 cm (c) 134° (d) 84 cm^2
(3) (a) 8.94 cm (b) 17.9 cm^2 (d) 8.9 cm, 96°
(4) (a) 8.66 cm (c) 8.9 cm (d) 60°
(5) (a) 49 cm^2 (b) 9.90 cm (c) 38.5 cm^2 (d) 308 cm^2

Exercise 33a

(1) 0.4348
(2) 1.5013
(3) 0.8146
(4) 1.0926
(5) 1.983
(6) 4.493
(7) 0.3262
(8) 0.5585
(9) 0.0849
(10) 2.342
(11) 35° 20′
(12) 68° 10′
(13) 18° 40′
(14) 37° 42′
(15) 16° 27′
(16) 21° 19′
(17) 16° 5′
(18) 29° 25′
(19) 51° 17′
(20) 26° 17′
(21) 0.3934
(22) 0.2476
(23) 0.9886
(24) 0.6858
(25) 0.3681
(26) 0.6732
(27) 0.8420
(28) 0.7457
(29) 0.1688
(30) 0.8778
(31) 23°
(32) 42° 10′
(33) 67° 30′
(34) 26° 35′
(35) 68° 41′
(36) 37° 49′
(37) 77° 19′
(38) 16° 42′
(39) 13° 14′
(40) 41° 23′
(41) 0.4797
(42) 0.7254
(43) 0.8897
(44) 0.7677
(45) 0.4315
(46) 0.1199
(47) 0.9189
(48) 0.9461
(49) 0.6890
(50) 0.7833
(51) 20° 30′
(52) 43° 20′
(53) 64° 10′
(54) 73° 14′
(55) 26° 21′
(56) 54° 37′
(57) 61° 54′
(58) 32° 14′
(59) 61° 21′
(60) 43° 33′

Exercise 33b

(1) (a) 3.4 (b) 73° 37′ (c) 17.7 cm (d) 42.5 cm^2
(2) (a) 2.460 (b) 68° (c) 22° (d) 5.79 cm^2
(3) (a) 0.7143 (b) 45° 35′ (c) 44° 25′ (d) 4.9 cm
(4) (a) 0.3333 (b) 70° 32′ (c) 22.6 cm (d) 38° 56′
(5) (a) 13 cm (b) 0.4167, 22° 37′ (c) 0.3846, 67° 23′ (d) 0.75, 48° 35′

Exercise 34a

(1) {4, 9, 14, 19, 24}
(2) {4, 8, 12, 16, 20}
(3) {1, 3, 5, 7, 9}
(4) {2, 12, 22, 32, 42}
(5) T
(6) T
(7) F
(8) F
(9) T
(10) F
(11) T
(12) F
(13) F
(14) T
(15) F
(16) T
(17) T
(18) T
(19) T
(20) F
(21) F
(22) F
(23) T
(24) T
(25) F
(26) T
(27) T
(28) F
(29) F
(30) T
(31) 1
(32) 5
(33) 2
(34) 1
(35) 1
(36) 5
(37) 5
(38) 7
(39) 4
(40) 2
(41) 5
(42) 7
(43) 3
(44) 6
(45) 8

Exercise 34b

(1) (a)

$\oplus_3$	1	2	0
1	2	0	1
2	0	1	2
0	1	2	0

(b) 2
(c) 1, 0
(d) {2, 5, 8, 11, 14}

(2) (a)

$\oplus_4$	1	2	3	0
1	2	3	0	1
2	3	0	1	2
3	0	1	2	3
0	1	2	3	0

(b)

$\otimes_4$	1	2	3	0
1	1	2	3	0
2	2	0	2	0
3	3	2	1	0
0	0	0	0	0

(c) 2 (d) 1

(3) (a)

$\oplus_2$	1	0
1	0	1
0	1	0

(b)

$\otimes_2$	1	0
1	1	0
0	0	0

(c) {2, 4, 6, 8, 10} (d) {odd numbers}

(4) (a) {3, 9, 15, 21, 27} (b) 0 (c) 0 (d) {5, 11, 17, 23, 29}, prime, 25

(5) (a) 5 (b) 1 (c) 5 (d) 7

Exercise 35a

(1) 13.0
(2) 18.1
(3) 161
(4) 15.6
(5) 0.540
(6) 1.19
(7) 12.3
(8) 142
(9) 10.2
(10) 0.857
(11) 1.94
(12) 6.77
(13) 7.83
(14) 9.18
(15) 2.94
(16) 22.4
(17) 9.98
(18) 6.49
(19) 2.23
(20) 5.04
(21) 164
(22) 120
(23) 179
(24) 23.0
(25) 54.5
(26) 4.24
(27) 12.5
(28) 2.91
(29) 4.40
(30) 2.03

Exercise 35b

(1) (a) 63.4 cm^2 (b) 34.3 cm^3 (c) 307 g
(2) (a) 14.8 cm (b) 17.4 cm^2 (c) 210 cm^3 (d) 2 210 g
(3) (a) 14.4 cm^2 (b) 5.13 cm^3 (c) 100g
(4) (a) 8.70 cm (b) 22.7 cm^2 (c) 4.76 cm
(5) (a) 3.57 cm (b) 6.32 cm (c) 8.94 cm

Exercise 36a

(1) 5, 1
(2) 2, 1
(3) 2, 2
(4) 3, -1
(5) 1, 2
(6) 1, 1
(7) 2, 2
(8) 3, 1
(9) 1, 2
(10) 5, -1
(11) -3, 4
(12) -1, -2
(13) 1, $\frac{1}{2}$
(14) $\frac{1}{4}$, 1
(15) -2, 0
(16) $\frac{1}{2}$, $\frac{1}{2}$
(17) 3, $-\frac{1}{2}$
(18) 0, $\frac{2}{3}$
(19) $-\frac{1}{4}$, $\frac{3}{4}$
(20) $\frac{1}{3}$, $\frac{1}{3}$
(21) 8, 2
(22) 7 yr, 8 yr
(23) 8, 6
(24) 7, 6
(25) 75, 425

Exercise 36b

(1) (a) 9 (b) 1, 2 (c) 13 yr, 35 yr
(2) (a) -8 (b) -3, 1 (c) 10, -5
(3) (a) 17 (b) $\frac{1}{2}$, -1 (c) 3, 1
(4) (a) 22 (b) $-\frac{1}{2}$, 2 (c) 12, 8, no
(5) (a) no (b) 2, 1 (c) 10p, 15p

Exercise 37a

(1) 125
(2) 90
(3) £54
(4) £1.05
(5) 9.9 m
(6) 10.5 cm
(7) 81 kg
(8) 108 g
(9) 40 days
(10) 1
(11) £46
(12) £570
(13) 615
(14) 21p
(15) 12.5 l
(16) £87
(17) £2.40
(18) 3.72 m
(19) 54 kg
(20) £15
(21) 12 white, 10 black
(22) 20 boys, 15 girls
(23) 45 pass, 18 fail
(24) 35 men, 49 women
(25) 54°, 36°, 90°
(26) 21 maths, 12 science
(27) 12 black, 14 white
(28) 170 km
(29) 48p
(30) 55p

(31) £35
(32) 45
(33) 342 cm
(34) 4 500
(35) $9\frac{1}{2}$ hours
(36) 80
(37) 150
(38) 270 km
(39) 54 secs
(40) 3 hours 20 mins.

Exercise 37b

(1) (a) 10 days (b) 10 days (c) 6 days
(2) (a) 294 men, 318 women (b) 510 (c) 595
(3) (a) 21 (b) £3.78 (c) 14 (d) 28
(4) (a) 490 km (b) £13.31 (c) 10 hours (d) 700 km
(5) (a) 24 g/m^2 (b) 60p (c) 125 m^2

Exercise 38a

(1) 9.5
(2) 4.1
(3) 7.5
(4) 8.3
(5) 6.2
(6) 6.5
(7) 9.5
(8) 9.4
(9) 4.1
(10) 7.0
(11) 21
(12) 29
(13) 20
(14) 51
(15) 10
(16) 20
(17) 32
(18) 27
(19) 12
(20) 45
(21) 710
(22) 860
(23) 12
(24) 490
(25) 900
(26) 1 100
(27) 46
(28) 6 800
(29) 22
(30) 110
(31) 3.6
(32) 4.6
(33) 1.7
(34) 8.4
(35) 9.3
(36) 0.56
(37) 3.5
(38) 0.59
(39) 4.5
(40) 5.7
(41) 14
(42) 290
(43) 39
(44) 31
(45) 42
(46) 41
(47) 32
(48) 460
(49) 2 000
(50) 350

Exercise 38b

(1) (a) 26 (b) 0.45 (c) 12
(2) (a) 33 cm (b) 43 cm (c) 18 cm
(3) (a) 830 (b) 0.31 (c) 19 (d) 1.7
(4) (a) 29 (b) 52 (c) 84 (d) 89
(5) (a) 15 cm (b) 72 cm^2 (c) 58 cm^3

Exercise 39a

(1) F
(2) T
(3) F
(4) T
(5) T
(6) F
(7) T
(8) F
(9) F
(10) T
(11) T
(12) F
(13) F
(14) T
(15) T
(16) T
(17) T
(18) F
(19) F
(20) F
(21) (i) 30° (ii) 110° (iii) 70°
(22) (i) 90° (ii) 50° (iii) 50°
(23) (i) 60° (ii) 120° (iii) 120°
(24) Isosceles trapezium
(25) Kite
(26) 5.3 cm
(27) 3.6 cm
(28) 7.1 cm
(29) 3.5 cm

Exercise 39b

(1) (a) Rectangle (b) 45°, 135°, 135° (c) 8.0 cm
(2) (a) (i) T (ii) F (iii) F (c) kite
(3) (c) (i) F (ii) F (iii) T (d) rectangle
(4) (a) Rhombus (b) rhombus (c) 110°, 110° (d) 40 cm^2
(5) (a) {squares} (b) (i) T (ii) F (iii) T (c) (i) 14 cm^2 (ii) 6 cm^2 (iii) 20 cm^2

Exercise 40a

(1) $a(x + y)$
(2) $3(x + z)$
(3) $5(a + 2b)$
(4) $3(x - 2y)$
(5) $2b(1 - a)$
(6) $2(3x + 5y)$
(7) $3x(7y - 2x)$
(8) $2(3x - 1)$
(9) $2x(2 + 7x)$
(10) $5(3x - 1)$
(11) $(p + q)(r + s)$
(12) $(a + b)(c + d)$
(13) $(w + x)(y + z)$
(14) $(a - c)(b - d)$
(15) $(2x + 1)(x + y)$
(16) $(1 + x)(x + y)$
(17) $(2a + b)(c - d)$
(18) $(a + 2b)(c + 2d)$
(19) $(x + 3y)(2a - 1)$
(20) $(x + 1)(3y - 2)$
(21) $(x + 1)(x + 2)$
(22) $(x + 2)(x + 3)$
(23) $(x + 2)(x - 3)$
(24) $(x + 1)(x - 3)$
(25) $(x - 1)(x + 2)$
(26) $(x - 7)(x + 1)$
(27) $(x + 5)^2$
(28) $(x + 3)^3$
(29) $(x - 1)^2$
(30) $(x - 2)^2$
(31) $(3x + 1)(x + 1)$
(32) $(5x + 1)(x + 1)$
(33) $(2x + 1)(x - 3)$
(34) $(2x - 1)(x - 1)$
(35) $(2x + 3)(x + 1)$
(36) $(3x - 1)(x + 2)$
(37) $(2x + 1)(x + 3)$
(38) $(3x + 5)(x + 1)$
(39) $(2x + 1)(x + 7)$
(40) $(2x - 3)(x - 1)$
(41) $(2x + 1)(x + 6)$
(42) $(2x + 3)(x + 2)$
(43) $(3x - 2)(x + 2)$
(44) $(2x + 3)(x - 2)$
(45) $(3x + 1)(x + 4)$
(46) $(2x + 1)(x - 6)$
(47) $(3x + 2)(x + 5)$
(48) $(3x - 5)(x + 2)$
(49) $(3x + 2)(x - 2)$
(50) $(3x + 1)(x + 4)$
(51) $(a + b)(a - b)$
(52) $(x + y)(x - y)$
(53) $(x + 2y)(x - 2y)$
(54) $(a + 3b)(a - 3b)$
(55) $(2x - 3y)(2x + 3y)$
(56) $(3a + 2b)(3a - 2b)$
(57) $(4p + 3q)(4p - 3q)$
(58) $(2s + 5t)(2s - 5t)$
(59) $(5x + 2y)(5x - 2y)$
(60) $(7y + 8x)(7y - 8x)$

Exercise 40b

(1) (a) $3x(2x + 5y)$ (b) $(2p + q)(r - s)$ (c) $(3x - 2)(2x - 3)$ (d) $(3x + 7y)(3x - 7y)$
(2) (a) $7p(3q - 2a)$ (b) $(x + 3)(y + 2)$ (c) $(2x + 1)(x - 6)$ (d) $(5a + 9b)(5a - 9b)$
(3) (a) $(x^2 + x - 2)$ cm² (b) $(2x + 1)(3y - 2)$ (c) $(3x + 7)(x + 5)$ (d) $(4x + 5y)(4x - 5y)$
(4) (a) $2a(4p + q)$ (b) $(x + y)(x - y)$ (c) $(2x - 5)(x - 1)$ (d) $(9a + b)(9a - b)$
(5) (a) $(x + 2)$ cm (b) $(3x + 11)(x - 2)$ (c) $(a + b)$ cm (d) $(p + 5q)(p - 5q)$

Exercise 41a

(1) 15% profit
(2) 10% profit
(3) 20% profit
(4) 1% loss
(5) 4% profit
(6) 25% profit
(7) 22% loss
(8) 2½% loss
(9) 12½% loss
(10) 37½% profit
(11) £27
(12) £2
(13) £9
(14) £8
(15) £51.10
(16) £16.80
(17) 92p
(18) £9.40
(19) £4.50
(20) £6.80
(21) £69
(22) £47
(23) £74.40
(24) £26.25
(25) £68.75
(26) £4 400
(27) £43.74
(28) £13.80
(29) £22.41
(30) £19.25
(31) £121
(32) £220.50
(33) £252.15
(34) £324.48
(35) £1 144.90
(36) £96.80
(37) £636.54
(38) £302.50
(39) £648
(40) £811.20

Exercise 41b

(1) (a) 16% (b) £52.20 (c) £2.61
(2) (a) 20% (b) £42 (c) 8.4%
(3) (a) £137.50 (b) £110 (c) £159.50 (d) £160.38
(4) (a) £285 (b) £41 (c) £2.50
(5) (a) £66.50 (b) £30 (c) 4 years (d) £61.50

Exercise 42a

(1) $b = \frac{A}{l}$
(2) $l = \frac{V}{bh}$
(3) $a = \frac{A}{\pi b}$
(4) $r = \frac{A}{2\pi h}$
(5) $h = \frac{2A}{b}$
(6) $t = \frac{v - u}{a}$
(7) $l = \frac{P - 2b}{2}$
(8) $c = 2s - a - b$
(9) $x = \frac{P - 2y}{3}$
(10) $x = P(4y + 1)$
(11) $b = Q(a - c) - a$
(12) $p = \frac{q(X - 1)}{2}$
(13) $m = \frac{2K}{v^2}$
(14) $c = \sqrt{\frac{E}{m}}$
(15) $a = \sqrt{(c^2 - b^2)}$
(16) $t = \sqrt{\frac{2s}{a}}$
(17) $r = \sqrt{\frac{A}{4\pi}}$
(18) $A = \frac{d^2}{2}$
(19) $s = \frac{v^2 - u^2}{2a}$
(20) $V = \pi r^2 h$

Exercise 42b

(1) (a) 6.40 (b) $x^2 = r^2 - y^2$ (c) $x^2 = (r + y)(r - y)$ (d) 12.1
(2) (a) 57.7 cm^2 (b) $h = \frac{A}{2\pi r}$ (c) 2.39 cm
(3) (a) 314 cm^2 (b) $r = \sqrt{\frac{S}{4\pi}}$ (c) 2.0 cm
(4) (a) 600 (b) $P = \frac{100\,A}{100 + RT}$ (c) 750
(5) (a) 4 (b) $y = \frac{x^2}{Q}$ (c) $x = \sqrt{(Qy)}$ (d) 4.93

Exercise 43a

(1) 14 cm
(2) 9
(3) 15 cm
(4) 24 cm
(5) 16 times
(6) 5 times
(7) 50°, 60°, 70°
(8) 3 : 1
(9) 180 cm^2
(10) 20 cm
(11) 36 : 1
(12) 27 : 1
(13) 8 : 1
(14) 70 m^2
(15) 8 kg
(16) 70 cm^2
(17) ellipse
(18) yes
(19) 27 cm^2
(20) 9 cm^2

Exercise 43b

(1) (a) 6 cm (b) 9 m (c) 48 cm^2 (d) 900 m^3
(2) (a) 3 cm (b) 450 cm^2 (c) 30 cm^2 (d) 1 026 cm^3
(3) (a) 18 cm (b) 3 : 1 (c) 76° 45′ (d) 216 cm^3
(4) (a) 128 cm (b) 2 516 cm^2 (c) 16.8 l

Exercise 44a

(1) (i) 3 (ii) 4 (iii) 9
(2) (i) 15 (ii) 7 (iii) 0
(3) (i) 1 (ii) −9 (iii) 19
(4) (i) 8 (ii) 20 (iii) 11
(5) (i) 7 (ii) 49 (iii) 7
(11) (1, −2) max
(12) (−1, 2) max
(13) (−1, −5) max
(14) (1, −1) max
(15) (0, 5) min
(16) $(\frac{1}{2}, \frac{3}{4})$ max
(17) (−1, −10) max
(18) $(\frac{1}{2}, 2)$ min
(19) $(\frac{1}{2}, -2\frac{1}{2})$ min
(20) (1, −4) min

Exercise 44b

(1) (c) 12.25 cm^2 (d) 3.2 cm
(2) (c) (1, −4) min (d) 2
(3) (c) (−1, 2) max (d) −3
(4) (c) $(-\frac{1}{2}, -5\frac{1}{4})$ min (d) −1
(5) (c) −3 (d) 3

Exercise 45a

(1) 23.4 cm
(2) 20.6 cm
(3) 23.9 cm²
(4) 52.8 cm
(5) 93.2 cm²
(6) 71.8 cm²
(7) 41.8 cm²
(8) 40.5 cm²
(9) 13.1 cm²
(10) 45.4 cm²
(11) 11.7 cm
(12) 23.6 cm
(13) 51.8 cm²
(14) 7.59 cm
(15) 58.3 cm²
(16) 34.2 cm²
(17) 55.4 cm²
(18) 167 cm²
(19) 695 cm³
(20) 574 cm³
(21) 99.2 cm³
(22) 84.4 cm³
(23) 13.8 cm²
(24) 270 cm²
(25) 10.2 cm
(26) 57.3 cm²
(27) 160 cm³
(28) 91.6 cm²
(29) 96.5 cm²
(30) 1 840 cm³

Exercise 45b

(1) (a) 169 cm² (b) 7.50 cm (c) 149 cm³
(2) (a) 8.77 cm (b) 116 cm² (c) 55.4 cm² (d) 142 cm³
(3) (a) 414 cm² (b) 704 cm² (c) 1 410 cm³
(4) (a) 25.6 cm² (b) 4.92 cm (c) 19.6 cm²
(5) (a) 190 cm² (b) 285 cm² (c) 348 cm³ (d) 174 cm³

Exercise 46a

(3) (i) 4:1 (ii) 9:1 (iii) 1:25 (iv) 1:16
(4) (i) 3:1 (ii) 5:1 (iii) 1:4 (iv) 1:8
(5) (a) SSS (b) RHS (c) RHS (d) SSS (e) RHS (f) SSS (g) ASA (h) SSS (i) SAS

Exercise 46b

(1) (d) 4:1
(2) (a) 10 cm (c) 4:5 (d) 16:25
(3) (a) 49° (b) 9 cm (c) 4:9 (d) 6:5

Exercise 47a

(1) $\begin{pmatrix} 7 & 7 \\ 6 & 4 \\ 5 & 4 \end{pmatrix}$
(2) $\begin{pmatrix} 7 \\ 1 \\ 2 \\ 11 \end{pmatrix}$
(3) $\begin{pmatrix} 5 & 0 \\ 5 & 1 \end{pmatrix}$
(4) $\begin{pmatrix} 11 \\ 3 \end{pmatrix}$
(5) $\begin{pmatrix} 9 & 6 & 4 \\ 4 & 0 & -7 \end{pmatrix}$
(6) $\begin{pmatrix} -1 \\ -4 \end{pmatrix}$
(7) $\begin{pmatrix} 5 & -1 \\ -1 & 6 \end{pmatrix}$
(8) $\begin{pmatrix} -4 & 0 \\ -3 & 2 \end{pmatrix}$
(9) $\begin{pmatrix} -5 & 2 \\ 1 & -3 \end{pmatrix}$
(10) $\begin{pmatrix} 0 & 1 & -2 \\ 1 & 0 & -1 \end{pmatrix}$
(11) (15)
(12) (13)
(13) (9)
(14) (33)
(15) $\begin{pmatrix} 9 \\ 18 \end{pmatrix}$
(16) $\begin{pmatrix} 16 \\ 6 \end{pmatrix}$
(17) $\begin{pmatrix} 0 \\ 0 \end{pmatrix}$
(18) $\begin{pmatrix} 2 \\ 6 \end{pmatrix}$
(19) $\begin{pmatrix} 27 \\ 11 \end{pmatrix}$
(20) $\begin{pmatrix} 2 \\ 4 \end{pmatrix}$
(21) $\begin{pmatrix} 2 & -1 \\ 9 & 8 \end{pmatrix}$
(22) $\begin{pmatrix} 12 & 0 \\ -2 & 1 \end{pmatrix}$
(23) $\begin{pmatrix} 2 & -6 \\ -2 & 6 \end{pmatrix}$
(24) $\begin{pmatrix} 6 & 3 \\ 12 & 9 \end{pmatrix}$
(25) $\begin{pmatrix} 0 & 0 \\ 0 & 0 \end{pmatrix}$
(26) $\begin{pmatrix} 13 & 5 \\ 5 & 2 \end{pmatrix}$
(27) $\begin{pmatrix} 1 & 0 \\ 0 & 1 \end{pmatrix}$
(28) $\begin{pmatrix} 5 & 4 \\ 6 & 5 \end{pmatrix}$
(29) $\begin{pmatrix} -4 & 10 \\ 2 & 2 \end{pmatrix}$
(30) $\begin{pmatrix} -6 & 9 \\ 6 & -6 \end{pmatrix}$

Exercise 47b

(1) (a) $\begin{pmatrix} 4 & 6 \\ 2 & 14 \end{pmatrix}$ (b) $\begin{pmatrix} 7 & 5 \\ 2 & 6 \end{pmatrix}$ (c) $\begin{pmatrix} 13 & 1 \\ 12 & -5 \end{pmatrix}$ (d) $\begin{pmatrix} 27 & 8 \\ 4 & 3 \end{pmatrix}$

(2) (c) reflection (d) $\mathbf{M}^2 = \begin{pmatrix} 1 & 0 \\ 0 & 1 \end{pmatrix}$

(3) (c) enlargement

(4) (b) reflection (c) rotation

(5) (a) rectangle (c) shear (d) 2 square units

Exercise 48b

(1) (c) 104 cm (d) 73°
(2) (c) 85°
(3) (d) 5.6 cm
(4) (d) 13.4 cm
(5) (d) 70° (e) 54°

Exercise 49a

(1) 66%
(2) 18.0 km/l
(3) 5.5
(4) 166 cm
(5) £5.40
(6) 58%
(7) 2
(8) 47 km/h
(9) 46%
(10) 87.8 km/l
(11) 2.8
(12) 12.5 cm
(13) 67.4 kg
(14) 567 h
(15) 8.26 g
(16) $\frac{1}{6} = 0.17$
(17) $\frac{1}{4} = 0.25$
(18) $\frac{1}{13} = 0.08$
(19) $\frac{3}{10} = 0.30$
(20) $\frac{1}{2} = 0.50$
(21) $\frac{1}{2} = 0.50$
(22) $\frac{3}{13} = 0.23$
(23) $\frac{1}{5} = 0.20$
(24) $\frac{1}{4} = 0.25$
(25) $\frac{1}{52} = 0.02$

Exercise 49b

(1) (a) 289 (b) 5.78 (c) 20% (d) 0.2
(2) (a) 36% (b) $\frac{9}{25} = 0.36$ (c) 4.24 (d) 56%
(3) (a) $\frac{3}{10} = 0.3$ (b) $\frac{1}{4} = 0.25$ (c) $\frac{3}{4} = 0.75$
(4) (a) £819.26 (b) £1 319.85 (c) £36.88
(5) (a) $\frac{1}{2} = 0.5$ (b) $\frac{11}{20} = 0.55$ (c) $\frac{1}{20} = 0.05$ (d) $\frac{7}{10} = 0.7$

Exercise 50a

(9) 4.1 cm
(12) 4.0 cm
(13) 5.4 cm
(14) 5.0 cm
(15) 2.3 cm
(16) 2.2 cm
(17) 4.3 cm

Exercise 50b

(1) (b) kite (c) 8.0 cm
(2) (b) rhombus (c) 6.8 cm (d) 45.2 cm²
(3) (a) 5.1 cm (c) △'s XYP and YZQ, △'s XYQ and XZP (d) 41 cm²
(4) (a) 4.1 cm (b) 4.5 cm (c) 7.5 cm

Exercise 51a

(1) $-1, 2$
(2) $-1, -3$
(3) $1, 1$
(4) $1, -5$
(5) $-1, 7$
(6) $-2, 3$
(7) $2, 5$
(8) $-3, 7$
(9) $5, -3$
(10) $-4, 2$
(11) $-3, 2$
(12) $3, 2$
(13) $-2, -3$
(14) $6, -1$
(15) $1, 8$
(16) $-2, -2$
(17) $-\frac{1}{3}, 2$
(18) $3, 3$
(19) $-\frac{1}{5}, 3$
(20) $\frac{1}{3}, -3$
(21) $-\frac{1}{2}, 1$
(22) $\frac{1}{2}, -3$
(23) $-\frac{2}{3}, 1$
(24) $-\frac{1}{7}, 2$
(25) $\frac{2}{11}, -1$
(26) $\frac{3}{2}, -1$
(27) $\frac{2}{3}, 1$
(28) $-\frac{3}{2}, 2$
(29) $\frac{3}{2}, -5$
(30) $\frac{2}{3}, -2$
(31) $1, 2$
(32) $0.5, -1$

(33) $-0.5, 2$
(34) $0.2, -1$
(35) $1.5, -2$
(36) $0.5, -0.5$
(37) $2.5, -2$
(38) $-1.5, 1$
(39) $0.5, 0.5$
(40) $-1, 2$

Exercise 51b

(1) (a) $1\frac{1}{2}$ (b) 1, 7 (c) $\frac{2}{3}, -5$
(2) (a) $(x - 2)(x + 1)$ cm² (b) $x^2 - x - 6 = 0, x = 3$ cm (c) 4.1 cm
(3) (a) $4x + 4$ (b) $x^2 + x$ (c) $x^2 - x - 2 = 0$ (d) $R = 2\frac{1}{2}$
(4) (a) $(2, -1)$ min (b) -4 (c) 1, 3
(5) (a) (1 1) max (b) -2 (c) 0, 2

Exercise 52a

(1) ∠A = 32° 46′, ∠B = 57° 14′, c = 8.68 cm
(2) ∠A = 78° 30′, ∠B = 11° 30′, c = 12.1 cm
(3) ∠A = 47° 10′, ∠B = 42° 50′, b = 9.30 cm
(4) ∠A = 27° 11′, ∠B = 62° 49′, b = 8.90 cm
(5) ∠A = 48° 7′, ∠B = 41° 53′, a = 26.2 cm
(6) ∠B = 65°, a = 3.54 cm, c = 8.38 cm
(7) ∠B = 50°, a = 7.13 cm, c = 11.1 cm
(8) ∠B = 28°, b = 6.44 cm, a = 9.73 cm
(9) ∠A = 8°, b = 6.04 cm, a = 0.854 cm
(10) ∠A = 47°, b = 7.64, a = 8.19 cm
(11) 11 m
(12) 18.5 m
(13) 904 m
(14) 63.5 m
(15) 50° 40′
(16) 6.67 m
(17) 48 m
(18) 69° 27′, 69° 27′, 41° 6′
(19) 108 km, N 21° 48′ E
(20) 65 km, 239° 37′
(21) 3.8 m
(22) 3.15 m
(23) 3.34 m
(24) 9.0 cm
(25) 49° 48′, 65° 6′, 65° 6′
(26) 59.9 km
(27) 115 km
(28) 55.6 km
(29) $43\sqrt{2}$, 60.8 cm
(30) $8\sqrt{3}$, 13.9 cm

Exercise 52b

(1) (a) 101°, 48° (b) 3.76 cm (c) 18.1 cm² (d) 4.26 cm
(2) (a) 93 km (b) 207 km (c) 227 km (d) 335° 47′
(3) (a) 8 cm (b) 32 cm² (c) 8π cm²
(4) (a) 39° (b) 55 m (c) 68 m
(5) (a) 43° 9′ (b) 93° 42′ (c) 6.58 cm (d) 21.6 cm²

Exercise 53a

(1) 13.2 cm
(2) 28° 41′
(3) 106° 16′
(4) 32°
(5) 64°
(6) 24°
(7) 66°
(8) 90°
(9) 131°
(10) 50°
(11) 6 cm
(12) 4 cm
(13) 16 cm
(14) 8 cm
(15) 146°
(16) 90°
(17) 60°
(18) 45°
(19) 106°
(20) 44° 25′
(21) 2.5 cm
(22) 102°
(23) 42°
(24) 81°
(25) 8 cm

Exercise 53b

(1) (a) 27° 23′ (b) 54° 46′, 27° 23′ (c) 62° 37′ (d) 6.66 cm
(2) (a) 60° (b) 60° (c) 70° (d) 50°
(3) (a) 8 cm (b) 9 cm (c) 28° 4′ (d) 30° 58′
(4) (a) 60° (b) 60° (c) 30°
(5) (a) 4.8 cm (b) 25° (c) 66° (d) 41°

Exercise 54a

(1) 60 km/h
(2) 0.09 m/h
(3) 72 km/h
(4) 140 km/h
(5) 1 575 km
(6) 231 km
(7) 375 m
(8) 231 km
(9) $2\frac{1}{2}$ h
(10) 2 h 5 min
(11) $4\frac{1}{2}$ h
(12) 5 days
(13) 51 km/h
(14) 27.5 km/h
(15) 40 km/h
(16) 400 km/h
(17) 44 km/h
(21) 10.15 a.m., 100 km from Leeds
(22) 1.30 p.m., 90 km from London
(23) 7 m/s
(24) 8 m/s
(25) 48 km/h

Exercise 54b

(1) (a) 4 h (b) 53.75 km/h (c) 27 l (d) 8.0 km/l
(2) (b) Brighton (c) 40 km/h
(3) (b) 08.24, 24 km from Preston (c) 40 km/h
(4) (b) 7.5 m/s (c) 10 m/s
(5) (b) 15 m/s (c) 10 m/s

Exercise 55a

(1) 940 km, 046°
(2) 6.1 km, 295°
(3) 2.3 km, 152°
(4) 87 km, 217°
(5) 3.4 km, N 35° W
(6) 182 km/h, 327°, 7°
(7) 183 km/h, 171°, 11°
(8) 193 km/h, 301°, 9°
(9) 272 km/h, 263°, 8°
(10) 265 km/h, 330°, 10°
(11) 5 kg, 100°
(12) 15 kg, 161°
(13) 3.4 kg, 242°
(14) 130 kg, 176°
(15) 130 kg, 155°

Exercise 55b

(1) (b) 4.7 km (c) 309° (d) 14 mins
(2) (a) 160 km/h (b) 360° (c) 9°
(3) (b) 11° (c) 38 km/h (d) 112°
(4) (b) 6.7 km/h (c) $63\frac{1}{2}$° (d) $26\frac{1}{2}$°

Index